AF248541

SOLID-STATE
DC MOTOR DRIVES

MONOGRAPHS IN MODERN ELECTRICAL TECHNOLOGY

ALEXANDER KUSKO, SERIES EDITOR

1. Solid-State DC Motor Drives by Alexander Kusko

SOLID-STATE
DC MOTOR DRIVES

ALEXANDER KUSKO

The M.I.T. Press
Cambridge, Massachusetts, and London, England

PREFACE

The family of solid-state devices termed thyristors, of which the silicon-controlled rectifier (SCR) is the most prominent, represents the most important development in the control of electric power in the past ten years. The impact of the thyristor on the field of dc motor drives has been particularly great because the combination of the thyristor circuit and the dc motor has resulted in a happy economic and technical marriage. Drives using the combination are found from 1/60 hp to 10,000 hp and in such diverse applications as trigger controlled electric drills, printing presses and steel rolling mills.

Technical material on solid-state dc drives is found in the technical literature, in conference proceedings, in manuals published by manufacturers, and in catalogs. Some of the technology is based on grid-controlled rectifier theory, some on magnetic-amplifier techniques, and some on conventional electric machine control principles. On the other hand, some of the technology is new and made possible by the size, cost, and characteristics of these thyristors.

The purpose of this book is to bring together the technology of solid-state dc drives for the reader who wants to understand the principles, to specify and purchase drives, or to design his own drives. The book is not a design text; design material is available in manufacturers' manuals. The material in the book stems from the author's consulting and teaching experience and from previously published papers. References are given at the end of each chapter for the reader who wishes to pursue the subject in more detail.

Motor drives using solid-state power control elements are still in their early stages of commercial development. The next twenty-year period will see the electric automobile with thyristor speed control, the vast expansion of urban transportation using electric motor driven and thyristor-controlled electric trains, heavy on- and off-highway vehicles of all types using gas turbines and electric drives, and new generations of air-conditioners, refrigerators and other home appliances using solid-state controlled motors of all kinds. The principles used in these drives will not differ from those in this book, although the packaging and ratings may change.

Cambridge, Massachusetts   ALEXANDER KUSKO

TABLE OF CONTENTS

Preface   v

1   Drive Applications   1

    1.1   Adjustable Speed Drives   1
    1.2   Constant-Horsepower Drives   2
    1.3   Series Universal Motor Drive   4
    1.4   Auxiliary Functions   4
    1.5   Methods for Driving DC Motors   5
    1.6   Torque-Speed Diagrams   6
    1.7   Summary   9

2   DC Motor Characteristics   10

    2.1   Shunt Motor   10
    2.2   Dynamics   12
    2.3   Series Motor   16
    2.4   Motor Parameters   20
    2.5   Commutation   24
    2.6   Summary   27

3   Rectifier Operation   28

    3.1   Single-Phase Rectifier–Resistance Load   28
    3.2   Single-Phase Rectifier–Reactive Load   31
    3.3   Three-Phase Rectifier–Resistance Load   37
    3.4   Summary   42

4   Single-Phase Drives   44

    4.1   Half-Wave Drives   44
    4.2   Full-Wave Drives   48
    4.3   Gaudet Drive Circuit   51
    4.4   Discontinuous Conduction   53
    4.5   Commercial Aspects   55
    4.6   Summary   55

5   Three-Phase Drives   58

    5.1   Incomplete Bridge   58
    5.2   Complete Bridge   61
    5.3   Dynamics   64
    5.4   Commercial Equipment   66
    5.5   Summary   68

6   Series Universal Motor Drives   70

    6.1   Basic Circuits   70
    6.2   Operation   73
    6.3   Analysis, Phase Control–Unsaturated   76
    6.4   Analysis, Phase Control–Saturated   79
    6.5   Phase Control–Interpretation   80
    6.6   Summary   82

7   DC-DC Drives   84

    7.1   Application   84
    7.2   Principle   84
    7.3   Chopper Circuit   86
    7.4   Regeneration   89
    7.5   Summary   91

8   Specification of Drives   93

    8.1   Speed   93
    8.2   Reversing–Braking   95
    8.3   Acceleration   96
    8.4   Torque-Current Limit   96
    8.5   Equipment   97

9   Control Equipment   99

    9.1   Firing Circuits   99
    9.2   Speed Sensing   104
    9.3   Current Sensing   105
    9.4   Motor Matching   107
    9.5   Construction   110

9.6   Field Winding Supply Methods   116
9.7   Speed Control By Field Weakening   118
9.8   Protection   119

Table of Symbols   123

Index   125

1 DRIVE APPLICATIONS

Drives are characterized by the shape of the speed-torque curves they produce and by the types of motors used to provide the mechanical power. They are also described by the horsepower size, e.g. fractional or integral horsepower, and by the supply, e.g. single-phase or three-phase. The latter descriptions apply to the physical nature of the equipment rather than to the function or performance characteristics.

1.1 *Adjustable Speed Drives*

The bulk of the solid-state dc motor drives operate under adjustable speed control. They utilize dc motors with independent field supplies and obtain speed control by control of the armature voltage. For any one control setting, the motor is made to operate at substantially constant speed from zero to maximum torque load by utilizing some type of sensing and feedback circuit. A family of typical speed-torque curves is shown in Figure 1.1

The ratings of commercial adjustable speed drives range from 1/60 to 3 hp for single-phase fractional and low-integral horsepower motors; three-phase drives are built from about one hp to 500 hp in standard packages and up to 10,000 hp as special designs for mill drives. The ranges of speed adjustment are typically six-to-one for standard units and up to 30-to-one for special units. The speed regulation for fractional-horsepower drives is typically five per cent droop of rated speed from no load to full load, and one per cent for the larger horsepower sizes.

The circuits consist of three portions regardless of size. The armature voltage is supplied from the ac line through a single-phase or three-phase thyristor controlled rectifier assembly. The field current is supplied from an auxiliary controlled or noncontrolled rectifier. A control system measures the speed of the motor, compares it with the reference or control setting, and generates the gating signals for the armature thyristor rectifier to maintain the speed. Further features, such as reversing contactors, acceleration limit and dynamic braking, are incorporated as needed.

The typical family of speed-torque curves shown in Figure 1.1 has a constant speed region for each control setting and a torque-limit region. The purpose for the torque-limit region is to prevent the control system

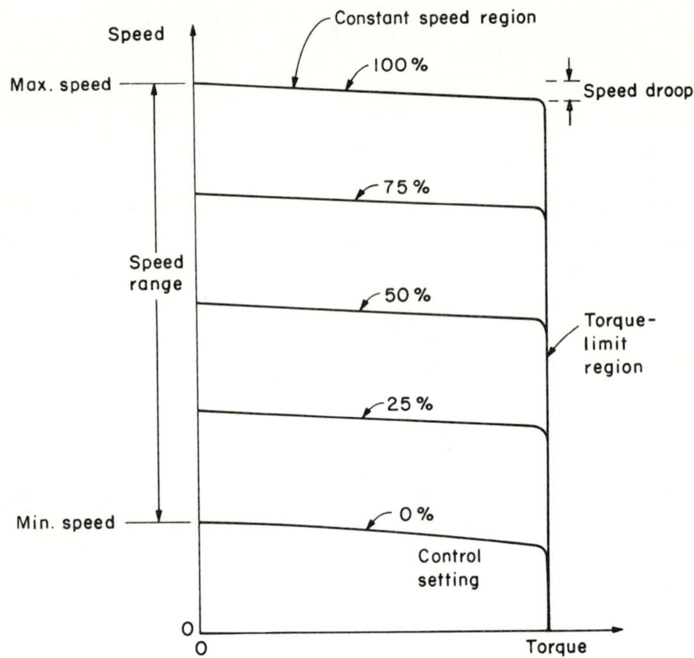

Figure 1.1   Speed-torque characteristics of adjustable speed drive.

from maintaining constant speed beyond the thermal capability of the motor and the thyristors, that is, overloading the motor and the rectifiers. The dc motor does not have the torque-limit region as an inherent characteristic; the control system must provide it. The region is also referred to as current limit and acceleration limit.

1.2 *Constant-Horsepower Drives*

Whereas the adjustable speed drives described above usually operate from ac lines having a power capability much greater than the drive itself, drives such as are used in vehicles are limited to power outputs set by the vehicle prime mover. Such drive systems are capable of delivering more mechanical power than the prime mover can develop and must be constrained by the control system to the horsepower available. These drives are known as constant-horsepower drives.

The requirement for a constant-horsepower drive is not met by the inherent open-loop characteristics of any of the standard dc motors with their external controls. The speed-torque characteristics of the series and shunt motor superimposed on the constant-horsepower restriction are shown in Figure 1.2. Both types of motors will operate outside the restriction unless they are restrained by a control system.

The series motor with adjustable series resistance R_a, whose characteristics are shown in Figure 1.2a, approximates a constant-horsepower drive for certain values of resistance R_a. For this reason, the series motor drive is sometimes referred to as a constant-horsepower drive. Practically, dc series motor traction drives for diesel-electric locomotives and off-highway vehicles are made to reflect constant-horsepower loading on the prime mover by making relatively small corrections to the system as a function of speed.

The shunt motor with adjustable armature voltage V_a can be made to operate along a constant-horsepower line as shown in Figure 1.2b by continuously controlling the voltage V_a as a function of speed and torque. The control operation requires major changes of voltage V_a compared to minor changes of resistance R_a for the series motor.

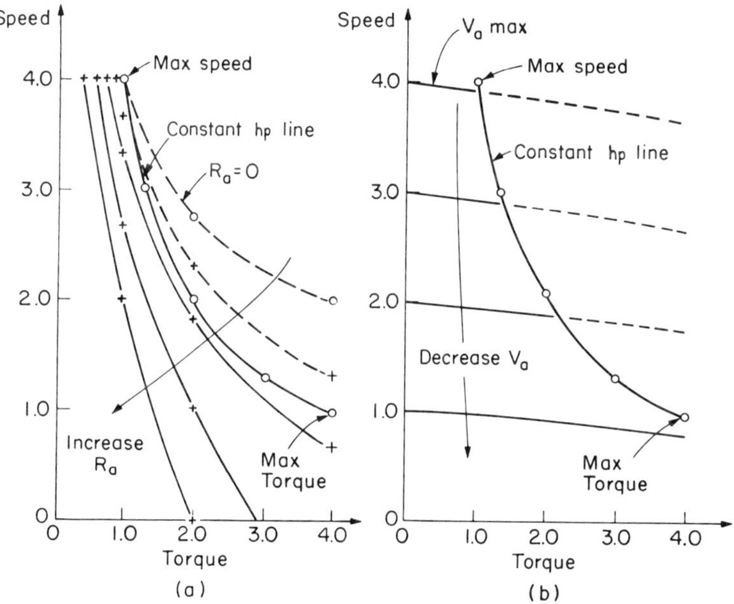

Figure 1.2   Constant-horsepower drive requirement as satisfied by:
(a) series motor with adjustable resistance R_a and
(b) shunt motor with adjustable armature voltage V_a.

The principal problem of building a constant-horsepower drive resides in the wide range of speed and torque over which the motor must operate. The maximum motor torque is set at the lowest speed; the maximum motor speed is set at the lowest torque. Thus the motor and its solid-state converters must have a horsepower rating of k times the value of the constant horsepower, where k = the steady-state speed range. This is true for *any* motor that is used for the constant-horsepower drive.

1.3 *Series Universal Motor Drive*

The series universal motor has a special place in drive applications as a compact, high-speed source of power for portable tools, appliances, and other uses. The motor can be speed controlled by operating it from an ac line and employing a half-wave or full-wave thyristor with phase control for adjusting the voltage applied to the motor. The controls are usually built into the motor or tool housing and are operated by a trigger or a knob on the tool.

The circuits used to control the series universal motor are simple and inexpensive. They are either of the nonfeedback or feedback type, depending upon the application. The simplest type consists of a half-wave thyristor with an *RC* firing circuit for phase control. The control element, such as the trigger, varies directly the resistance of the firing circuit. The next improvement is the use of a full-wave thyristor with the same type of control. Finally, the most complicated of these circuits involves speed feedback to keep the motor speed from drooping with load. The armature voltage serves as the measure of speed.

The speed versus load torque characteristics of a series universal motor used in an electric drill are shown in Figure 6.9. The curves are plotted as a function of thyristor firing angle α. The effect of retarding the firing angle is to reduce the voltage applied to the motor and to squeeze the current into a shorter conduction angle, both of which cause the speed to droop. When the motor losses are added to the load torque it is seen that the curves approach constant-torque shape. The use of feedback tends to reduce the droop, but a better solution would be to use a shunt motor to obtain constant-speed operation.

1.4 *Auxiliary Functions*

The speed regulating accuracy of the dc drive is dependent upon the means used for sensing and measuring the speed. The standard accuracy drives use the armature voltage of the dc motor either directly or compensated for the *IR* drop as a speed signal. The accuracy can be improved by using a tachometer for a speed signal. Arbitrarily high accuracy can be obtained by using digital speed measuring and reference systems, although adjustable-frequency ac synchronous-motor drives are more suitable for every high accuracy.

Bidirectional operation requires a reversing contactor for solid-state dc drives because the thyristor bridges are single ended. Drives for servo operation that must operate around standstill and for either direction are built with two sets of thyristor bridges or power transistor amplifiers, one for each direction of operation.

Acceleration of a dc motor is produced by the application of the air gap torque to the inertia and damping of the motor and its load. The acceleration must be controlled to protect the load and to limit the current carried by the motor and the thyristors, as well as the supply line. The control is effected by measuring the armature current with a resistor or transductor and overriding the speed control signal during the acceleration period.

Deceleration of the solid-state dc drive requires extracting the kinetic energy from the inertia of the motor and its load and either dissipating it or returning it to the line. Systems using motor-generator sets to provide the armature voltage for the motor automatically return energy to the line because they operate bilaterally. However, solid-state drives must either use a resistor which is switched across the armature to dissipate the energy, or must use a thryristor bridge which will operate as a line-commutated inverter and pump the energy back into the line.

1.5   *Methods for Driving dc Motors*

Drive systems for dc motors fall into three categories: 1, those suitable for operation from a fixed-voltage dc bus; 2, those using ac-dc motor generator sets; 3, those operating from a fixed-voltage ac line using static (nonrotating) rectifier means.

Fixed-voltage dc supply systems are found in railways, industrial plants, ships, and on some aircraft. Shunt motors are supplied from such systems with the armature directly connected and the field winding connected through a field rheostat. The rheostat is used to vary the field current and obtain speed variations up to about three-to-one. The motor is capable of constant-horsepower loading because the torque capability decreases as the speed increases. At any rheostat setting the speed remains within a typical five per cent band from no load to full load.

Series motors are used with dc supply systems to obtain wide speed and torque ranges where constant-speed operation is not required. Resistance control elements connected in series with the motor provide speed and torque control under the manual operation of the operator. Thyristor chopper circuits produce similar characteristics without the power loss of the resistors.

Prior to the extensive development of solid-state devices, and even today, motor-generator sets are used to provide dc power from an ac line for dc motor drives. The armature of the dc motor is supplied directly from the main generator while the field winding of the motor is supplied from an auxiliary generator or rectifier. The armature voltage of the motor is adjusted by adjusting the generator field current from full-voltage positive to full-voltage negative. Hence, the dc motor can be reversed easily and automatically regenerates power to the ac line through the motor-generator set when speed reduction is commanded. The cost, losses, and space required are the drawbacks of the motor-generator set system, compared to the advantages of reversing, regenerative braking, smooth current waveforms, and large short-time overload capability.

Static systems for providing dc power from an ac line include magnetic amplifiers, saturable reactors, gas-discharge rectifiers and thyristors. Each element of the system consists of a rectifier means and a control means. The control means all operate by delayed conduction either by magnetic voltage blocking or by phase control of the controlled rectifier element. The basic systems are all unidirectional and require additional equipment for reversing and braking.

The solid-state dc motor drive systems described in this book offer advantages of relatively low cost, small physical size, and wide ranges of control. In addition the drive systems are commercially available for all standard line voltages, horsepowers, and regulation requirements. However, for certain applications, other-than-solid-state drives may represent better engineering solutions. Motor-generator sets are very useful for reversing applications; resistance-controlled series motors are difficult to surpass for varying-speed applications where dc power is available.

1.6 *Torque-Speed Diagrams*

The operation of drives for a change in speed or torque is usually presented on a four-quadrant chart. The horizontal axes are designated plus and minus motor torque applied to the load; the vertical axes are designated plus and minus speed. Steady-state forward operation takes place in the first quadrant; steady-state reverse in the third quadrant. Four typical conditions are shown in Figure 1.3.

A simple speed reduction is shown in Figure 1.3a. The motor is initially driving the load at speed N_1. The controller is shifted to call for speed N_2. The controller drops the armature source voltage to make the motor slow down. If the source is a unidirectional thyristor bridge with no dynamic braking resistor, the armature current and motor torque become zero and the motor slows down by virtue of the load. When the motor speed reaches speed N_2, current flows and the torque is reestablished. However, if the drive is provided with dynamic or regenerative braking, a reverse armature current flows, producing a braking torque, and the motor more quickly reaches the speed N_2.

Figure 1.3b shows the behavior for an increase of speed. When the controller is advanced so that the armature source voltage is forced higher than the armature voltage, a large current and a large accelerating current can be produced. Usually, the drive has a current-limit circuit which holds the current and torque at a preset value during the accelerating period, until the speed reaches N_2.

Reversing the motor from speed N_1 forward to N_2 reverse is shown in Figure 1.3c. The command to reverse would normally produce large negative armature current as the armature source voltage is reversed. However, the current or torque limit whose function must reverse with the command maintains the current at a preset value as the motor comes to a stop in the second quadrant and accelerates to its reverse speed N_2 in the third quadrant.

Lowering a load or unwinding a reel is shown in Figure 1.3d for two types of load behavior. The load is initially being raised at speed N_1, then it is stopped and held at zero speed with some torque less than the hoisting torque. Finally, the controller is set to lower the load at a speed N_2. If the load has more weight than the friction of the drive, the load will tend to accelerate downward; the motor will have to apply positive torque to maintain the required negative speed N_2. On the other hand, if the weight is less than the friction, such as an "empty hook," or is balanced out, as in an elevator, the motor will have to apply a negative torque to drive the load downward.

The four-quadrant diagrams are useful for determining how the reversing, braking, and current-limit circuits must operate as the drive carries out its operating cycle.

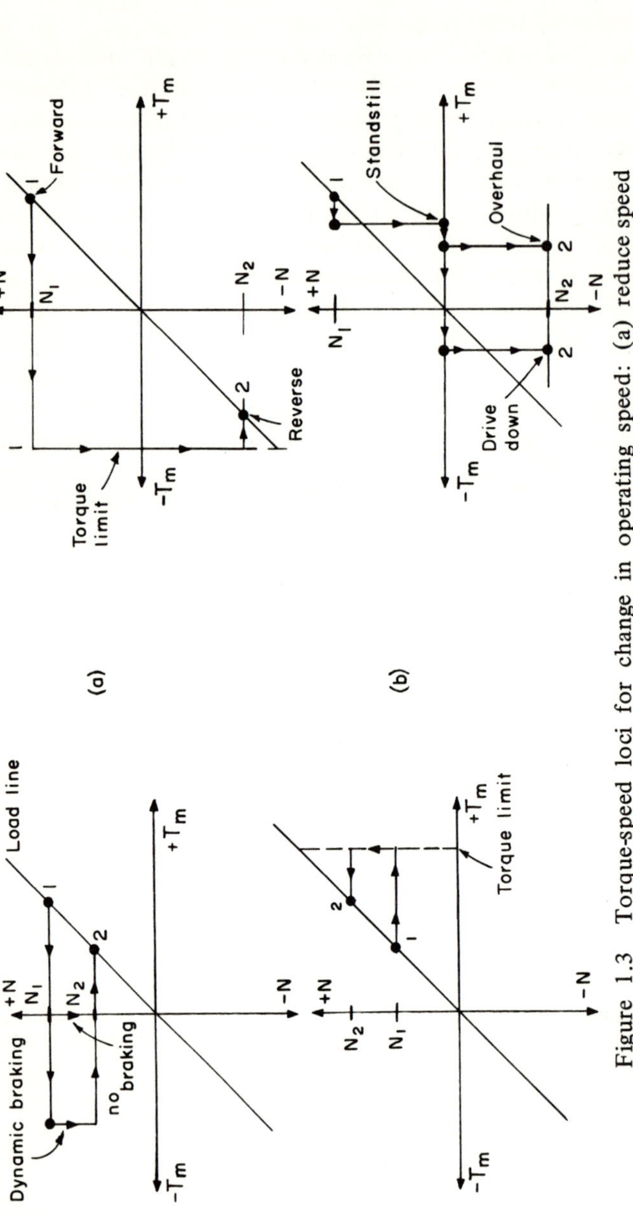

Figure 1.3 Torque-speed loci for change in operating speed: (a) reduce speed from N_1 to N_2 with inertia load; (b) increase speed from N_1 to N_2, torque limited; (c) reversing from N_1 to N_2 under torque limit; (d) lowering a load at speed N_2 under overhauling and drive conditions.

1.7 *Summary*

Solid state dc drives include a wide range of power ratings, torque-speed characteristics, and applications. The majority operate from ac sources and use the solid-state devices for rectification and control. The features are compactness, efficiency and lower cost than previously used methods.

The general theory and terminology of electric drives is treated in the classical book by Shoults, Rife, and Johnson[1] and more recently by Jones.[2] Electric drive applications to specific fields such as steel mills, machine tools, and transportation are described in the technical publications and conference proceedings of the Institute of Electrical and Electronic Engineers (IEEE). Technical material on design and performance of thyristor drive circuits is given by manufacturers of the devices.[3,4]

References

1. D. R. Shoults, C. J. Rife, and T. C. Johnson, *Electric Motors in Industry.* New York: John Wiley & Sons, Inc., 1942.
2. R. W. Jones, *Electric Control Systems.* Third edition. New York: John Wiley & Sons, Inc., 1953.
3. *SCR Manual.* Fourth edition. Auburn, N.Y.: General Electric Co., 1967.
4. *Silicon Controlled Rectifier Designers Handbook.* First edition. Youngwood, Pa.: Westinghouse Electric Corporation, 1964.

2 DC MOTOR CHARACTERISTICS

The flexibility of the dc motor for speed control, its overload capability, and its reliability make it the dominant means of providing a controllable source of mechanical rotating power in industry. The steady-state characteristics of the motor determine how the motor can be controlled; the dynamic characteristics determine how the control system must be designed to obtain the required response.

2.1 *Shunt Motor*

The shunt motor is operated in drive systems with the main or shunt field supplied with field current independently of the armature. The field current may be kept fixed or adjusted to supplement the speed range. A sketch showing the cross section of a shunt motor is given in Figure 2.1a. All of the possible windings are shown, although some may not be used in every application. The theory of operation of the shunt motor is given in any standard text, such as Fitzgerald and Kingsley.[1]

The shunt field winding establishes the basic magnetic field in the air gap under the poles, which reacts with the armature current to produce

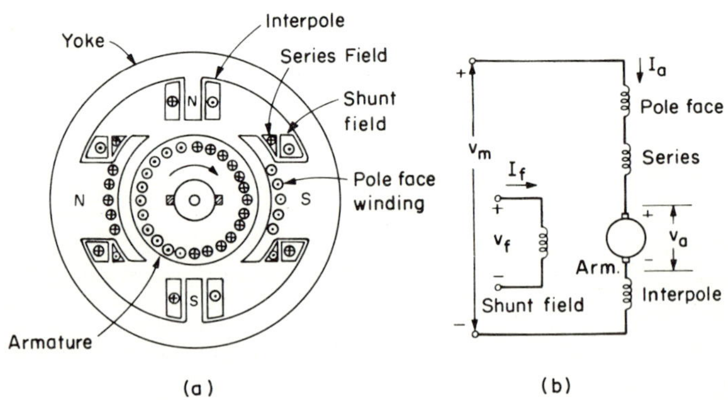

Figure 2.1 (a) Cross section, windings, and (b) circuit of dc shunt motor.

torque, and with the moving armature conductors to produce the generated armature voltage. The series field winding, used in some motors, acts to increase the speed droop when connected additively to the shunt field, and to decrease the droop for opposite connection. The interpole winding in the cross-field axis acts to assist commutation. Finally, the pole-face windings act to reduce the armature inductance and to prevent the armature current from distorting the magnetic field produced by the shunt field winding. The windings are connected as shown in Figure 2.1b.

The steady-state operation of the motor is governed by three equations. The total armature circuit voltage is

$$V_m = V_a + I_a R_a.$$ (2.1)

The generated armature voltage is given by

$$V_a = K_a \Phi_f N.$$ (2.2)

And, the internal torque is

$$T = K_t \Phi_f I_a.$$ (2.3)

The torque constant K_t and armature voltage constant K_a are equal in a consistent set of units (mks). The simultaneous solution of the three equations yields for the speed

$$N = \frac{V_m - T(R_a/K_t \Phi_f)}{(K_a \Phi_f)}.$$ (2.4)

The second term of the numerator of Equation 2.4 is usually small, say five per cent of the first term, V_m.

Equation 2.4 shows that the speed of the motor can be controlled in three ways. They are: 1, by the armature circuit voltage V_m which is nearly proportional to speed; 2, by the magnetic field Φ_f, which is inversely proportional to speed; 3, by the armature circuit resistance R_a, which is proportional to the speed droop.

The control of speed by the armature circuit voltage V_m, abbreviated armature voltage control, is the most desirable because the magnetic field can remain at full amplitude and full torque can be developed. Field control is accomplished by reducing the shunt field current and is used to extend the speed above the value for full armature voltage. The available torque decreases. Finally, armature circuit resistance control is not practical except in very small motors because of the power dissipation.

The characteristics for a full range of speed are shown in Figure 2.2. Armature voltage control raises the speed to N_1; field control raises it additionally to speed N_2. The available torque at rated armature current is

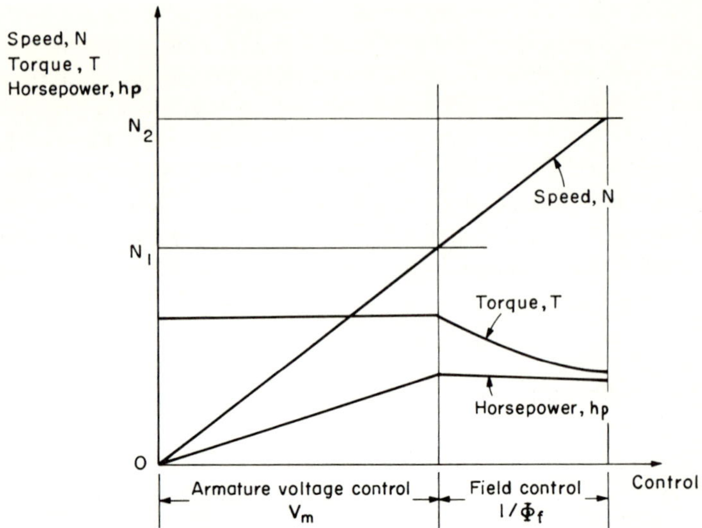

Figure 2.2   Speed control range of shunt motor under armature voltage and field control.

maximum to speed N_1 and drops with the field to speed N_2. The horsepower increases linearly to speed N_1 and remains constant to speed N_2.

The generated armature voltage V_a of Equation 2.2 is synonymous with the terms counter electromotive force, abbreviated CEMF and back emf, used in the drive industry. Equation 2.2 shows that the voltage V_a can be used as a measure of the motor speed N provided that the field magnetic flux Φ_f is constant. Most drive systems either use the voltage V_m of Equation 2.1, or subtract from it the $I_a R_a$ voltage, for a motor speed signal. Otherwise, a tachometer must be installed on the motor shaft to measure speed.

2.2 Dynamics

The dynamic behavior of the motor is governed by its energy storage properties. Energy is stored in two places: by magnetic fields in the magnetic circuit and by mechanical velocity in the inertia of the armature. Of course, energy is also stored in the inertia of the load and in magnetic fields of the electrical sources and must be included in the consideration of the dynamics.

The energy storage property of the magnetic fields is measured by inductance. As can be seen from Figure 2.1a, the armature and field circuits have windings but the magnetic axes are 90° from each other and the inductances theoretically have no mutual component. Hence, the armature inductance L_a, which includes the interpole and pole-face windings, can be considered independently of the field inductance L_f.

The dynamic equivalent circuit of the motor is shown in Figure 2.3.

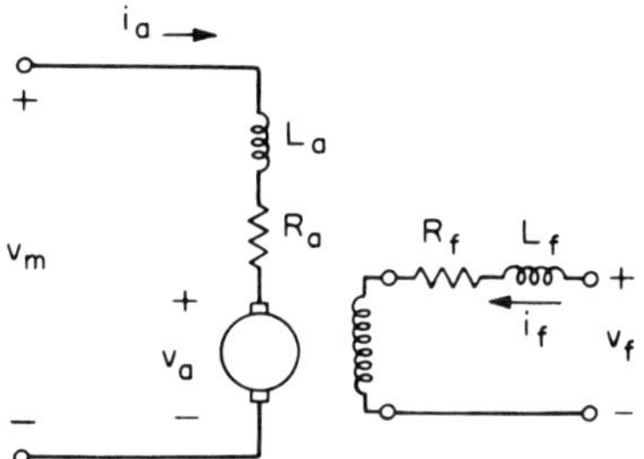

Figure 2.3   Dynamic equivalent circuit of motor.

The resistance R_a includes all of the windings in the armature circuit. The armature itself is the generator of the voltage v_a and the torque T. The mechanical parameters are not shown.

The differential equations which describe the dynamic behavior for armature voltage control at constant field are, for the armature circuit

$$v_m = i_a R_a + L_a \frac{di_a}{dt} + v_a, \qquad (2.5a)$$

$$v_a = K_a'\omega. \qquad (2.5b)$$

For the mechanical system

$$T = \omega B + J \frac{d\omega}{dt}, \qquad (2.6)$$

$$T = K_t i_a. \qquad (2.7)$$

The motor load is idealized as a torque proportional to velocity ω.

The dynamic properties are conventionally shown by means of a block diagram which shows the relationship of the differential equations for small amplitudes of the variables. The variables used are the transforms of the variables in the differential equations. The block diagram for the motor is shown in Figure 2.4.

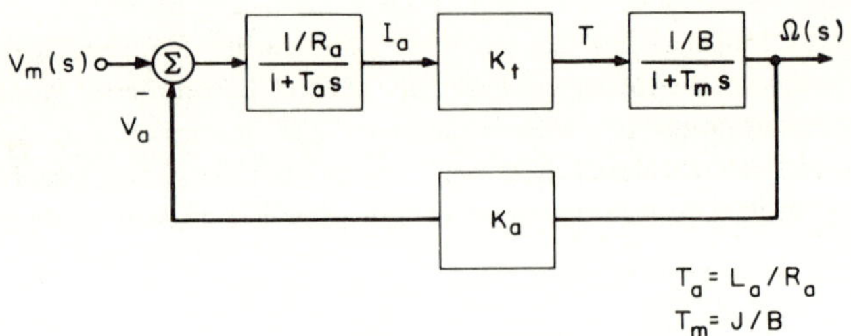

Figure 2.4   Block diagram of motor.

The relationship shown by the block diagram is expressed by a transfer function for the motor as

$$\frac{\Omega(s)}{V_m(s)} = \frac{K_t/R_a B}{(1 + T_a s)(1 + T_m s) + K_t K_a/R_a B}.$$

(2.8)

For the usual case of $(K_t K_a/R_a B) \gg 1$, the transfer function becomes

$$\frac{\Omega(s)}{V_m(s)} = \frac{1/K_a}{s^2/\omega_0{}^2 + 2\zeta s/\omega_0 + 1},$$

(2.9)

where

$$\omega_0{}^2 = \left(\frac{K_t K_a}{R_a B}\right)\left(\frac{1}{T_a T_m}\right) = \frac{K_t K_a}{L_a J},$$

(2.10)

$$\zeta = \frac{\omega_0}{2} \frac{T_a + T_m}{K_t K_a/R_a B}.$$

(2.11)

For large motors, the time constant $T_a < T_m$, and the transfer function becomes

$$\frac{\Omega(s)}{V_m(s)} = \frac{1/K_a}{1 + T_m' s},$$

(2.12)

where

$$T_m' = \frac{J R_a}{K_t K_a}.$$

(2.13)

It must be kept in mind that the dynamic parameters ω_0, J, and T_m' just described apply to small signal operation. For larger signals, such as during acceleration periods, the control system usually limits the motor current and torque so that the effective response time constant is governed only by the mechanical parameters. Likewise, during deceleration, the armature circuit may behave as though open, or may have braking resistance inserted, so that the response will not be described by the small-signal relationships. The small signal expressions are useful for determining the response to small variations of control and for stabilizing the system against oscillation.

Another useful method for studying the behavior of the dc shunt motor under transient conditions is to use the analog electric circuit of the mechanical portion. Using the analogs

$$T = k_1 i_1, \tag{2.14}$$

$$\omega = \frac{1}{k_1} v_1, \tag{2.15}$$

and the Equation 2.6, we obtain the equation for the analog electric circuit

$$i_1 = \frac{B}{k_1^2} v_1 + \frac{J}{k_1^2} \frac{dv_1}{dt}, \tag{2.16}$$

$$i_1 = Gv_1 + C \frac{dv_1}{dt},$$

where

$$G = \frac{B}{k_1^2}, \qquad C = \frac{J}{k_1^2}. \tag{2.17}$$

The analog of the mechanical system consists of a parallel GC circuit, where G is analogous to the damping B, and C is analogous to the inertia J. If the scale factor k_1 is made equal to K_a or K_t, then according to Equations 2.5 and 2.7 the current $i_1 = i_a$ and the voltage $v_1 = v_a$. The complete armature circuit can then be represented as shown in Figure 2.5.

The analog circuit in which the armature inertia acts to the electrical source as a capacitor is useful in visualizing the response of the motor to pulses of armature current from a rectifier and behavior under acceleration and deceleration.

The values of the equivalent capacitance seen by the source are large by electrical standards. For example, a motor having $J = 100$ kg $\cdot$ m^2, and $K_a = 2.5$ V $\cdot$ s/rad, has an analog or equivalent capacitance of

$$C = \frac{J}{K_a^2} = \frac{100}{6.25} = 16 \text{ F.} \tag{2.18}$$

A motor of that size could have $B = 5$ N $\cdot$ m $\cdot$ s/rad, for an equivalent conductance of

$$G = \frac{B}{K_a^2} = \frac{5}{6.25} = 0.8 \text{ mho.} \tag{2.19}$$

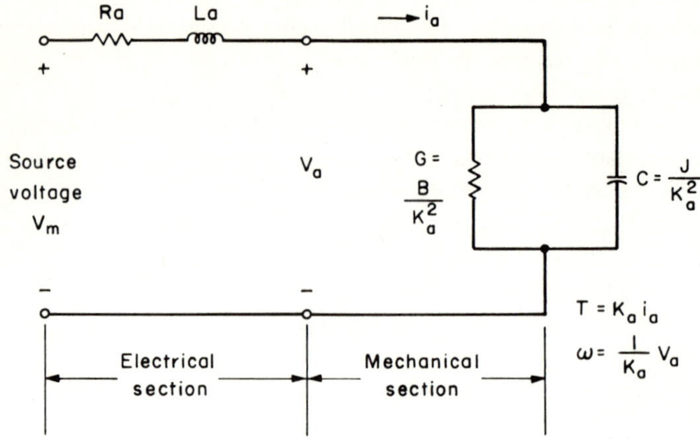

Figure 2.5   Electric circuit analog of the armature of a dc motor.

The strictly mechanical time constant of the motor would be $C/G = 20$ s. The motor would require almost two minutes to coast to a stop from full load at full speed.

2.3 Series Motor

The steady-state characteristics of the large series motor, over a few horsepower, are of interest for either adjustable voltage control, or series resistance control. The characteristics of the small series motor, less than one horsepower, are of interest for series thyristors operating under phase control. The large series motor can be considered as operating with the magnetic circuit not severely saturated. The small motor, since it is usually used in portable tools and the like for short periods at high loadings, is usually highly saturated.

The magnetic field of the large motor is assumed proportional to armature and series field current

$$\Phi_f = K_m I_a. \tag{2.20}$$

The current is then

$$I_a = \sqrt{T/K_m K_t}, \tag{2.21}$$

and the generated armature voltage

$$V_a = K_a N \sqrt{K_m T/K_t}. \tag{2.22}$$

Substituting in Equation 2.1, we obtain

$$V_m = K_a N \sqrt{K_m T/K_t} + R_a \sqrt{T/K_m K_t} = AN\sqrt{T} + BR_a\sqrt{T}. \tag{2.23}$$

Two conditions are of interest. First, if no external resistance is used, the second term of Equation 2.23 is small compared to the first term and the speed is given by

$$N \approx \frac{V_m}{A\sqrt{T}}.$$ (2.24)

The speed-torque curves represented by Equation 2.24 for three values of V_m are shown in Figure 2.6. The variation of V_m could be easily obtained from a thyristor bridge.

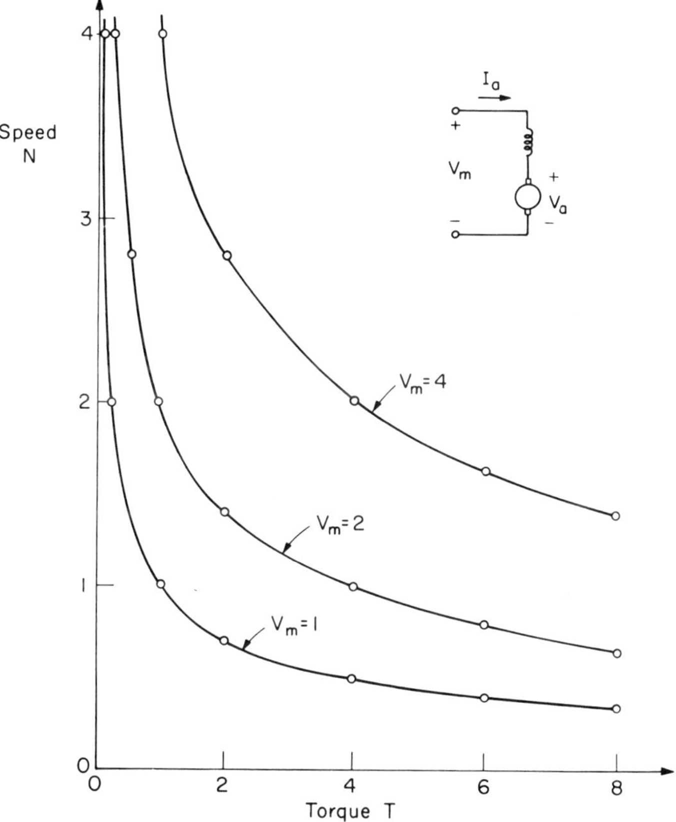

Figure 2.6   Speed-torque characteristics of series
motor under voltage control.

The second condition is that for added external resistance to R_a. The speed is then given by

$$N = \frac{V_m}{A\sqrt{T}} - \frac{B}{A} R_a.$$ (2.25)

The increments of resistance R_a produce fixed step reductions of speed all over the range. The effect is shown in Figure 2.7 for several values of resistance R_a.

The speed-torque curves of Figures 2.6 and 2.7 show that voltage control and resistance control of series motors are each more effective in different regions. Voltage control gives control at low values of torque,

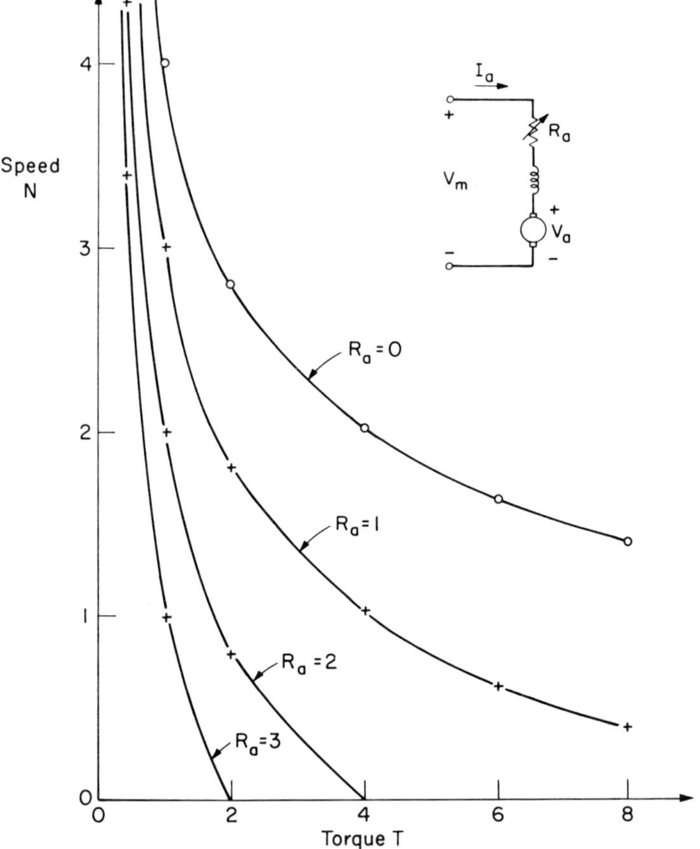

Figure 2.7   Speed-torque characteristics of series motor under series resistance control.

when the armature current is too small to be effective with resistance control. It also causes the motor to operate as a constant-speed source at low values of speed and voltage. Resistance control caused the motor to operate as a torque source at large values of resistance and produces finite values of stall torque. The equivalent effect of resistance control is obtained with some types of switching time-ratio solid-state controls.

The small series motor can be treated as though the magnetic circuit is saturated when current flows and the magnetic field is fixed at value Φ_s. The motor would tend to have speed-torque characteristics described by Equation 2.4 for the shunt motor. If resistance control of the speed is utilized, the curves will appear like those in Figure 2.8 for those torques

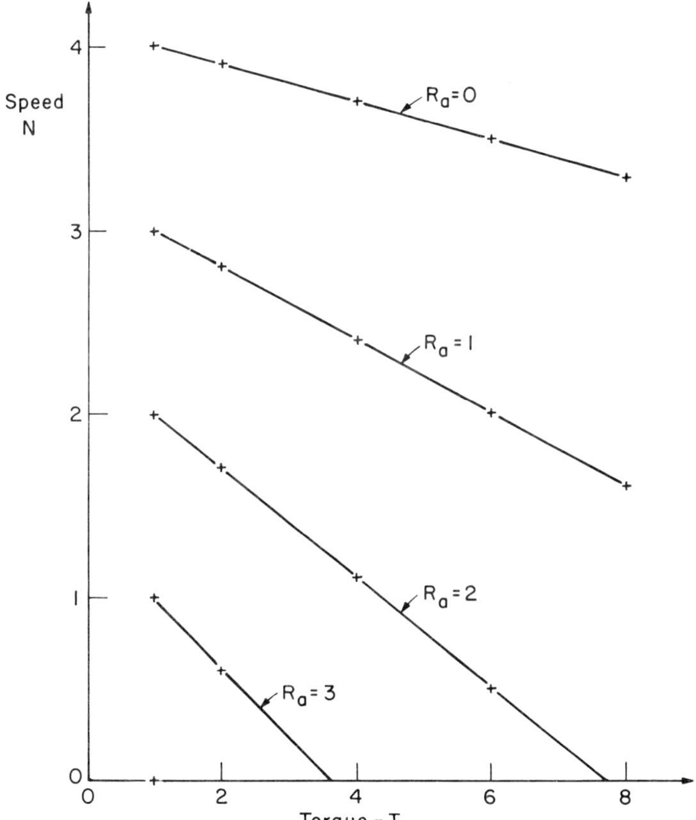

Figure 2.8   Speed-torque characteristics of saturated series motor under series resistance control.

corresponding to currents that saturate the structure. The series universal motor operating from phase controlled thyristors will be treated in more detail in Chapter 6.

2.4 *Motor Parameters*

The frame of the motor and the amount of active material basically determines the torque and cost of the motor. The speed at which the motor is operated determines the horsepower. Curves showing the size, weight, and cost of dc shunt motors as a function of horsepower are shown in Figure 2.9. The rated speed is a parameter on the curves. More specific information can always be obtained from manufacturers' catalogs.

A set of electrical and dynamic parameters for the General Electric line of Kinematic motors is given in Table 2.1. Such information is not provided by manufacturers and can be taken as typical for all low-inertia motors of the same rating.

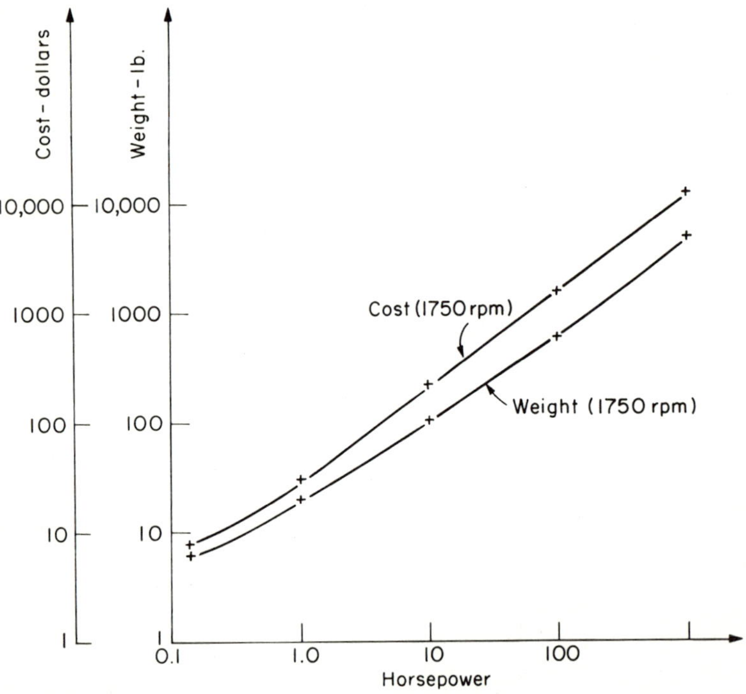

Figure 2.9 Weight and cost of typical dc shunt motors.

Additional types of motors are built to achieve special operation. Among them are the following:

1. *Printed Circuit Motors.* The armature consists of a circular insulating card upon which the radial armature conductors are printed. The motor has a very low electromechanical time constant and is used in wide bandwidth servos.
2. *Permanent Magnet Field Motors.* These are usually small motors in which the wound field is replaced by a permanent magnet structure. The purpose is to reduce cost and reduce losses.
3. *Torque Motors.* These are low-speed, large-diameter dc motors suitable for driving loads directly without gear trains. They can operate around standstill and require a reversible armature voltage power source.

4. *Split Field Motors.* These motors are usually used in reversible servo systems, either for field control, or as reversible series motors. They are usually less than one horsepower.

The subjects of matching dc motors to rectifiers at standard line voltages, commutation under rectifier operation, and effect of form factor on heating will be covered in Chapter 9.

Direct current motors for drive applications are built in a range of housings to fit the environment of the application. The electrical parameters and peak torque of a particular motor frame size are unaffected by the housing; however, the continuous or duty-cycle horsepower is affected because the heat dissipating ability is affected.

The basic types of housings are as follows:

1. *Drip-proof.* The interior of the motor is exposed to the ambient air. The ventilation openings are located on the underside of the motor. For operation down to the low-speed end of the speed range, the motor must be equipped with a blower to maintain forced ventilation.
2. *Totally Enclosed.* The interior of the motor is not exposed to the ambient air. Internal air within the motor is circulated by an internal shaft-mounted fan. All heat is exchanged through the motor shell.
3. *Totally Enclosed – Fan Cooled.* An external shaft-mounted fan is provided to blow air over the outside of the motor shell to improve the external heat transfer operation.
4. *Totally Enclosed – Dual Cooled.* Internal air is circulated through an external heat exchanger where heat is transferred to ambient air by an external motor-driven blower.

TABLE 2.1   Performance Data and Parameters of 1 to 150 hp High Performance Compensated 240-V Industrial dc Motors
(Courtesy Direct Current Motor and Generator Department, General Electric Company)

Performance Data

rpm	Motor base speed in revolutions per minute
R_{am}	Motor armature-circuit resistance in ohms at 25°C (X1.20 hot)
L_{am}	Motor armature-circuit inductance in henrys (unsaturated)
K_t	Motor torque constant in pound-foot per ampere
K_v	Motor voltage constant, CEMF, in volts/radian/second

Frame	rpm	R_{am}	L_{am}	K_t	K_v
283	3500	0.153	0.0013	0.35	0.605
	2500	0.301	0.0023	0.49	0.85
	1750	0.615	0.0045	0.7	1.21
	1150	1.426	0.0104	1.06	1.84
	850	2.608	0.0192	1.44	2.5
	500	7.56	0.055	2.45	4.23
	300	19.5	0.153	4.08	7.07
284	3500	0.142	0.0011	0.41	0.59
	2500	0.279	0.0021	0.57	0.82
	1750	0.570	0.0043	0.81	1.17
	1150	1.36	0.0100	1.23	1.78
	850	2.42	0.0185	1.67	2.42
	500	6.71	0.0532	2.75	3.98
	300	19.34	0.147	4.74	6.85
286	3500	0.070	0.00070	0.420	0.655
	2500	0.137	0.00140	0.588	0.917
	1750	0.280	0.00281	0.840	1.31
	1150	0.657	0.00650	1.23	1.98
	850	1.19	0.0120	1.73	2.69
	500	3.32	0.0344	2.86	3.99
	300	9.5	0.095	4.87	7.60
288	3500	0.045	0.00073	0.425	0.610
	2500	0.089	0.00144	0.600	0.850
	1750	0.180	0.00293	0.850	1.22

Frame	rpm	R_{am}	L_{am}	K_t	K_v
288	1150	0.415	0.00677	1.30	1.85
	850	0.762	0.0125	1.75	2.50
	500	2.21	0.0360	2.98	4.27
	300	6.1	1.00	4.90	7.10
365	3500	0.022	0.00055	0.42	0.63
	2500	0.041	0.0011	0.59	0.88
	1750	0.086	0.0022	0.84	1.26
	1150	0.199	0.0051	1.28	1.91
	850	0.368	0.0094	1.73	2.6
	500	1.06	0.027	2.94	4.42
	300	2.91	0.075	4.87	7.3
366	3500	0.0168	0.00026	0.43	0.64
	2500	0.0328	0.00050	0.612	0.896
	1750	0.067	0.0010	0.875	1.28
	1150	0.155	0.0024	1.33	1.95
	850	0.284	0.0044	1.7	2.56
	500	0.772	0.013	2.93	4.37
	300	2.27	0.035	5.1	7.25
367	2500	0.0203	0.00052	0.61	0.88
	1750	0.0415	0.0011	0.87	1.26
	1150	0.0963	0.0025	1.32	1.92
	850	0.176	0.0046	1.79	2.58
	500	0.478	0.013	2.98	4.3
	300	1.41	0.036	5.1	7.35

Frame	rpm	R_{am}	L_{am}	K_t	K_v
368	1750	0.0363	0.00085	0.87	1.26
	1150	0.0964	0.0020	1.32	1.92
	850	0.153	0.0036	1.79	2.60
	500	0.417	0.011	3.21	4.41
	300	1.24	0.29	5.07	7.33
503	1750	0.0144	0.0011	0.885	1.27
	1150	0.089	0.0025	1.33	1.95
	850	0.066	0.0045	1.93	2.77
	500	0.168	0.013	3.02	4.35
	300	0.500	0.036	5.13	7.4
504	1750	0.0100	0.00085	0.892	1.27
	1150	0.0237	0.0020	1.33	1.95
	850	0.0420	0.0036	1.82	2.6
	500	0.115	0.011	3.13	4.45
	300	0.342	0.029	5.2	7.4
505	1750	0.0099	0.00073	0.885	1.27
	1150	0.0206	0.0017	1.33	1.95
	850	0.0380	0.0031	1.81	2.6
	500	0.109	0.0090	3.1	4.45
	300	0.350	0.025	5.13	7.4
506	1150	0.0125	0.0012	1.33	1.95
	850	0.0230	0.0023	1.8	2.63
	500	0.0660	0.0065	3.06	4.47
	300	0.188	0.019	5.1	7.5

Nominal Performance Constants

T_{pu} 1.0 per unit torque in pound-feet (continuous torque, only of DP 1150-rpm or above or blower-ventilated)
W_f Power for full field in watts
J_m Motor inertia in pound-feet second^2
T_m Motor inertial time constant in seconds (JR_a/K_tK_v)
Cfm Forced air in cubic feet per minute
P Static pressure drop in inches of water
$1/T$ Bandwidth in radians per second (ω_o)

Frame	T_{pu}	W_f	J_m	T_m	$1/T$	Ventilation	
						Cfm	P
283	15	150	0.050	0.036	61	150	1.00
284	30	160	0.065	0.039	58	150	1.00
286	45	180	0.087	0.022	67	150	1.00
288	60	200	0.115	0.020	56	150	1.00
365	75	210	0.218	0.018	48	350	1.25
366	120	220	0.292	0.018	62	350	1.25
367	150	230	0.340	0.013	55	350	1.25
368	180	242	0.412	0.014	56	350	1.25
503	300	325	1.34	0.018	28	800	1.9
504	375	410	1.43	0.013	30	800	1.9
505	450	430	1.62	0.022	31	800	1.9
506	570	500	2.08	0.011	31	800	1.9

NOTE: For an application requirement, the horsepower rating and frame size can be chosen from the table. Considerations are ventilation, enclosure, continuous rms torque (or horsepower) and peak torque. Ventilation and enclosure affect the continuous rms torque capacitory of a given frame size.

The rms torque or the peak momentary overload torque may be the limiting requirement. Using the rated or 1.0 per torque (T_{pu}) for the frame size chosen for thermal rating, use the maximum momentary load curves of Figure A to identify the overload capability, peak torque $= T_{pu}x$ (per cent overload/100). If the peak torque capability is not sufficient, then a new frame size must be chosen, based on peak torque. The curves of Figure A are defined as follows:

1. Instantaneous loads are defined as 0.5 seconds duration or less repeated not oftener than once every minute.
2. Occasionally repeated loads are defined as 5 seconds duration or less repeated not oftener than once every 5 minutes.
3. Frequently repeated loads are defined as 1 minute duration or less repeated not oftener than once in a period 20 times the duration.
4. Curves apply regardless of whether speed is obtained by armature voltage or shunt field control.
5. Curves also apply for regenerating operations.

With the frame and base speed chosen, performance data can be taken from the table.

Horsepower Rating and Frame Sizes
Drip-proof (DP), 60° C Rise

hp	3500	2500	1750	1150	850	500	300
			Frame Size				
			Speed in rpm				
1	—	—	—	—	—	283	284
2	—	—	—	—	283	284	286
3	—	—	—	283	284	286	366
5	—	—	283	284	286	288	367
7½	—	283	284	286	288	366	368
10	—	283	284	286	288	367	503
15	283	284	286	365	366	368	504
20	284	286	288	366	367	503	505
25	286	288	365	366	368	504	—
30	286	288	366	367	368	505	—
40	288	366	366	368	503	506	—
50	—	366	367	503	504	—	—
60	—	367	368	503	505	—	—
75	—	—	503	504	505	—	—
100	—	—	503	505	—	—	—
125	—	—	504	506	—	—	—
150	—	—	505	—	—	—	—

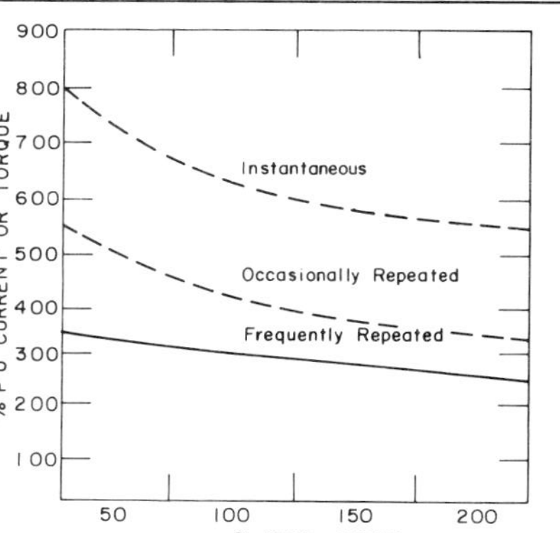

Figure A   Maximum momentary loads.

For a typical drive system having a 20-to-1 speed range, a drip-proof motor can be operated at rated torque down to 60 per cent speed and at 60 per cent torque down to 5 per cent speed. Forced ventilation must be used to operate at rated torque over the complete speed range

2.5 *Commutation*

Commutation is an ever-present problem in dc machine design and operation. The problem resolves itself simply into what conditions result in sparking at the brushes and in reduced life of commutator and brushes. Usually excessive sparking causes reduced life. It is generally known that commutation becomes poorer with heavier current, with higher speed, and with reduced main field flux. With operation from rectifier sources, it has been found that commutation becomes poorer with increased current ripple.

An effective method for placing a number on commutation is the black-band test. In the test, the current in the interpole winding is boosted and bucked from its operating value until sparking occurs. The limits of the band are then expressed in amperes or per cent of operating current. The wider the band, the better the commutation at the condition of test.

The result of a black-band test on a large dc motor is shown in Figure 2.10[2] for zero ripple and 6.2 per cent ripple, as a function of armature current. The bandwidth is seen to be reduced for the ripple and to be offset as the current increases, as though the ripple current tended to

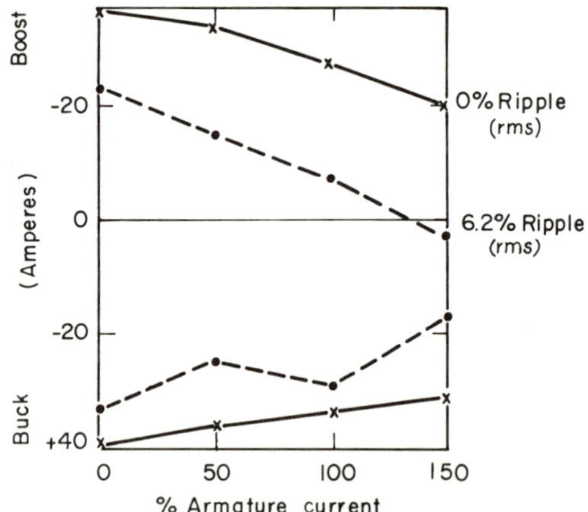

Figure 2.10   Black band of a 1250-kW, 720-rpm generator operating at 750 V with 0% and 6.2% rms ripple.

boost the interpole flux. A set of curves showing the bandwidth and divergence for a dc motor as a function of speed at constant current, no ripple, is shown in Figure 2.11.[3] The bandwidth is seen to decrease as the speed is raised, first by armature voltage control, then by field weakening. The curves are typical for four-pole wave-wound motors from 10 to 100 hp.

The typical effect of ripple on the net bandwidth is shown in Figure 2.12.[3] The net bandwidth is half the total bandwidth less the divergence.

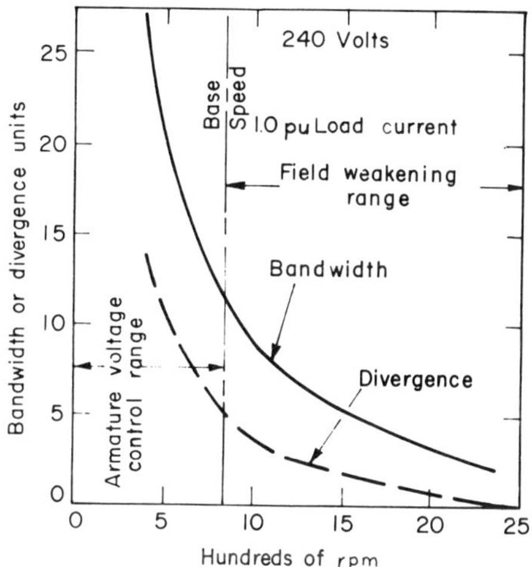

Figure 2.11   Constant current variable speed dc commutation characteristic showing bandwidth and divergence.

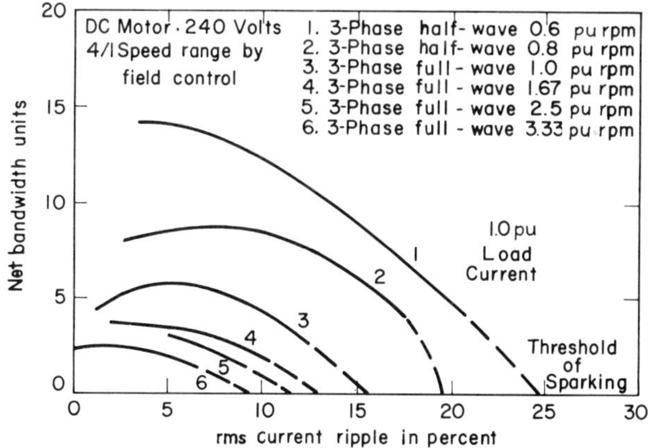

Figure 2.12   Typical bandwidth characteristic on rectified power.

As the ripple and speed increase, the net bandwidth decreases for satisfactory commutation. An application curve can be drawn for a machine as shown in Figure 2.13[3] for allowable ripple versus speed at 1.0 pu current. Speed is controlled by armature voltage to 1.0 pu speed and by field weakening above that point. The ripple characteristics of a typical power supply with several values of external armature inductance is shown superimposed on the motor curves.

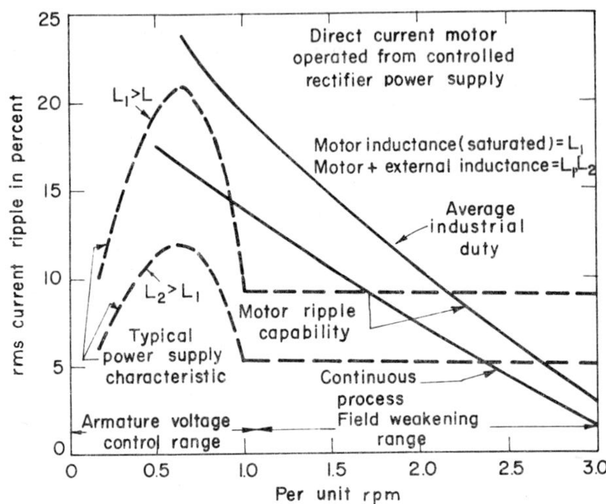

Figure 2.13 Coordination of rectified power supply characteristic and motor commutating capability curve as an aid to applying dc motors on rectified power.

The ripple current deteriorates commutation by at least two mechanisms. First, the eddy currents produced in the interpole and magnetic structure cause the interpole flux variations to lag behind the armature current, so that the flux and current are not matched. Second, the ripple component raises the reactance voltage in the commutated coil. The narrower the black band, the more vulnerable the machine to sparking with ripple.

The ripple can be reduced by utilizing more rectifier phases, by using an external armature choke, and by increasing by design the inherent armature inductance. If the machine is not required to operate over a wide speed range with field weakening, then the armature can be built larger, with smaller air gaps and a weaker main field to obtain higher armature inductance. An external choke can be designed to contribute its inductance at lower currents, where the ripple is high, and to saturate at higher currents, where the ripple is low.

2.6 Summary

With the exception of low-powered shunt and series motor drive circuits, solid-state speed control systems utilize feedback control loops to obtain prescribed motor output behavior. The speed-torque curves of the motors are then overcome by the gain of the speed control loops. The dynamic parameters of the motor affect the response of the closed-loop system.

Details on motor operation and analysis are presented in standard references on electric machinery, such as Fitzgerald and Kingsley.[1] References that show the relationship between design factors and dynamic parameters of motors are Lebenbaum,[4] and Newton and Rasche.[5]

References

1. A. E. Fitzgerald and C. F. Kingsley, Jr., *Electric Machinery*. Second edition. New York: McGraw-Hill Book Company, Inc., 1961.
2. R. M. Dunaiski, "The effect of rectifier power supply on large dc motors," *Trans. AIEE*, pt. III, vol. 79, pp.255–259, June 1960.
3. N. Kaufman, "An application guide for the use of dc motors on rectified power," *Trans. IEEE, Power App. Systems*, pp.1006–1009, October 1964.
4. P. Lebenbaum, Jr., "The design of dc motors for use in automatic control systems," *AIEE Trans.*, vol. 68, pp. 1089–1094, 1949.
5. G. C. Newton, Jr. and R. W. Rasche, "Can electric actuators meet missile requirements?" *AIEE Trans.*, vol. 80, pt. II, pp. 306–312, 1961.

3 RECTIFIER OPERATION

For speed control of a dc motor we must have a rectifier assembly which is capable of delivering adjustable levels of direct voltage to the armature, the field, or both, of the motor. The rectifier can operate from a single-phase or a three-phase line and must be able to tolerate the effects of the generated armature voltage and the load inductance. This chapter will deal with rectifier operation into passive loads of resistance, inductance, and capacitance; operation with motors will be described in Chapters 4, 5, and 6.

3.1 *Single-Phase Rectifier–Resistance Load*

The basic element of rectifier operation is the diode for noncontrolled operation, and the thyristor for controlled operation. For our purposes, the diode can be modeled as a switch which closes when the anode-to-cathode voltage is positive, and has a forward voltage drop of about one volt. The thyristor can be modeled in the same way with the added requirement that the gate terminal receive a pulse to become conducting.

The waveforms of the half-wave circuit with resistance load are shown in Figure 3.1. The load current and load voltage have the same waveform. The diode conducts during the positive half-cycles of line voltage, that is, $\omega t = 0$ to π, 2π to 3π, etc. The load voltage is described by an average value V_n, which is given by

$$V_n = \frac{1}{2\pi} \int_0^\pi \sqrt{2}\, V_0 \sin \omega t \, d(\omega t) = \frac{-\sqrt{2}\, V_0}{2\pi} \cos \omega t \Big]_0^\pi$$

$$= \frac{\sqrt{2}\, V_0}{\pi} = 0.45 V_0 . \tag{3.1}$$

The rms current I_n in the load is given by

$$I_n = \left[\frac{1}{2\pi} \int_0^\pi \frac{2 V_0^2}{R_n^2} \sin^2 \omega t \, d(\omega t) \right]^{1/2}$$

$$= \frac{V_0}{R_n} \left[\frac{1}{\pi} \left(\frac{\omega t}{2} + \frac{\sin^2 \omega t}{4} \right)_0^\pi \right]^{1/2} = 0.707 \frac{V_0}{R_n} . \tag{3.2}$$

Operation of the half-wave thyristor differs from the diode by the addition of the firing angle α as a variable. The expressions for the diode represent the maximum values obtainable with the thyristor at $\alpha = 0°$.

The circuit and waveforms of the half-wave thyristor operation are shown in Figure 3.2. The average load voltage V_n is now described in terms of the rms line voltage V_0 as

$$V_n = \frac{1}{2\pi} \int_{\alpha}^{\pi} \sqrt{2}\, V_0 \sin \omega t \, d(\omega t) = \frac{-\sqrt{2}}{2\pi} V_0 \cos \omega t \bigg]_{\alpha}^{\pi}$$

$$= \frac{\sqrt{2}\, V_0}{\pi} \frac{(1 + \cos \alpha)}{2}$$

$$= 0.225 V_0 (1 + \cos \alpha). \tag{3.3}$$

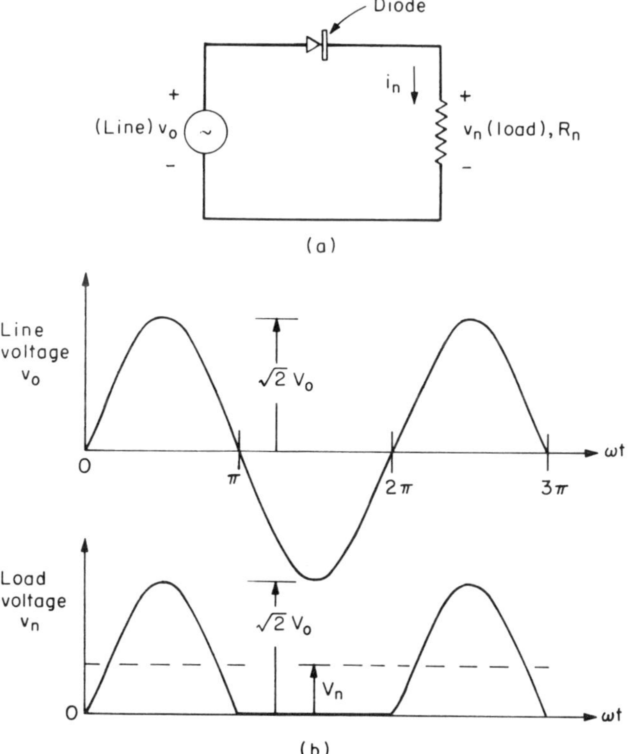

Figure 3.1   Operation of half-wave rectifier with resistance load.

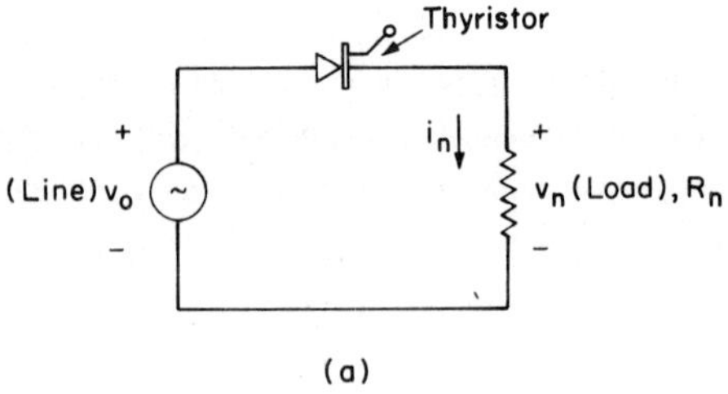

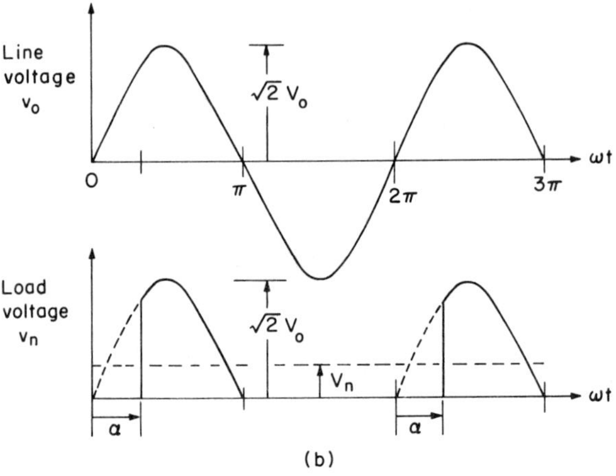

Figure 3.2   Operation of half-wave thyristor
with resistance load.

The rms current I_n is given as

$$I_n = \frac{V_0}{R_n}\left[\frac{1}{\pi}\left(\frac{\omega t}{2} + \frac{\sin^2 \omega t}{4}\right)_{\alpha}^{\pi}\right]^{1/2}$$

$$= \frac{V_0}{R_n}\left[\frac{1}{2} - \frac{\alpha}{2\pi} - \frac{\sin^2 \alpha}{4\pi}\right]^{1/2}. \tag{3.4}$$

The average voltage V_n and rms current I_n are shown in Figure 3.3 as a function of firing angle α.

3.2 *Single-Phase Rectifier–Reactive Load*

Two types of reactive load are used for modeling the actual loads imposed by dc machines on rectifiers. One is the series combination of inductance and resistance, as found in armature and field circuits. The second is the parallel combination of capacitance and resistance, as used to describe the electrical equivalent in the armature circuit of the mechanical inertia and damping. The current and voltage waveforms of each type of load will be described for both the diode and controlled rectifier.

The circuit and waveforms for the diode with series R and L load are shown in Figure 3.4. The current i_n is that for a transient phenomenon in each positive half-cycle as the diode applies a half-cycle pulse to the circuit. The response consists of a current that would be the steady-state ac value, called i_1, and a transient current i_2 that exists to satisfy the boundary conditions at $t = 0$ and decays with time.

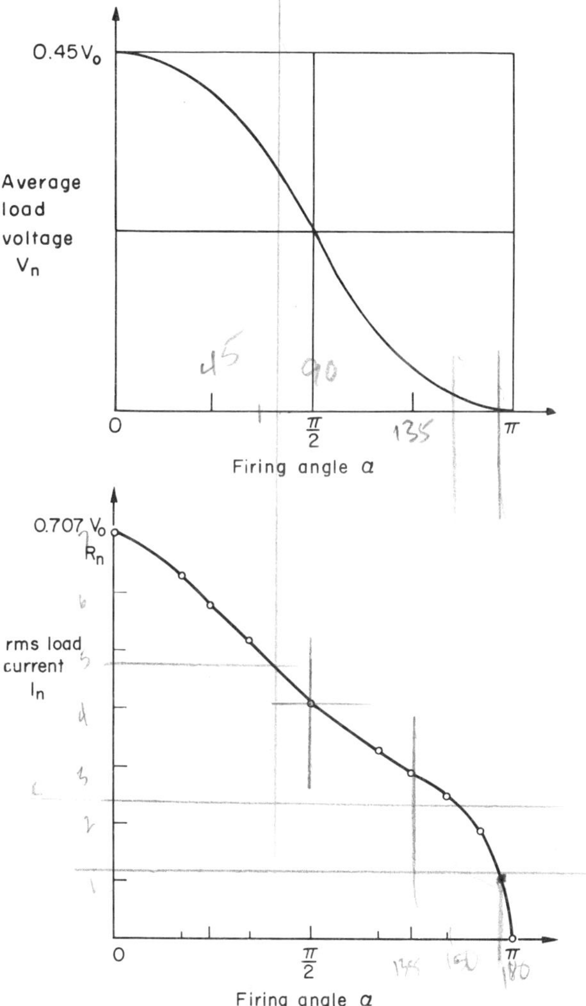

Figure 3.3   Plot of the average voltage V_n and rms current I_n as a function of firing angle.

The steady-state value is of the form

$$i_1(t) = \frac{\sqrt{2}\,V_0}{(R^2 + X^2)^{1/2}} \sin(\omega t - \phi), \qquad (3.5)$$

where

$$\phi = \arctan \frac{X}{R},$$

$$X = \omega L.$$

The transient current i_2 must force the net current $i_n(0) = 0$, hence has the form

$$i_2(t) = -I_1(0)e^{-t/\tau} \qquad (3.6)$$

where

$$i_1(0) = \frac{\sqrt{2}\,V_0}{(R^2 + X^2)^{1/2}} \sin(-\phi),$$

$$\tau = L/R.$$

The resultant current is thus

$$i_n(t) = i_1(t) + i_2(t), \qquad (3.7)$$

and exists from $t = 0$ to $\omega t = \beta$, the extinction angle, where the current $i_n(t) = 0$ and the diode blocks to end the half-cycle.

The example shown in Figure 3.4 is for $R = X = 1.0$ per unit. For an rms line voltage $V_0 = 1.0$ per unit, the two current components have the form

$$i_1(t) = 1 \sin(\omega t - \pi/4), \qquad (3.8)$$

$$i_2(t) = 0.707 e^{-\omega t}. \qquad (3.9)$$

The extinction angle is approximately $\beta = 5\pi/4 = 225°$. The peak current occurs at approximately $3\pi/4 = 135°$.

The inductance L cannot support any net average voltage over one period. Hence, the net shaded area in Figure 3.4, representing the inductance volt-time area, must always come out to zero when the current reaches zero at $\omega t = \beta$. Moreover, the average current I_n over a cycle is the average value of the line voltage v_0 from $\omega t = 0$ to $\omega t = \beta$ applied to the circuit resistance

$$I_n = \frac{1}{2\pi R} \int_0^\beta \sqrt{2}\,V_0 \sin \omega t \, d(\omega t). \qquad (3.10)$$

Hence, as the X/R ratio is increased, the amplitude of the current and its average value decreases, as β shifts toward 2π. At $\beta = 2\pi$, the current is zero over the cycle; the inductance never allows the current to become established.

For a thyristor instead of a diode, the current starts at the instant the thyristor is gated on, rather than at $\omega t = 0$. The waveforms for three values of firing angle α are shown in Figure 3.5 for $\omega L = R$. The expression for the instantaneous current i_n is made up of the same two terms as for the diode current, except that the transient current i_2 is evaluated at α.

The current i_n is given by

$$i_n(t) = i_1(t) + i_2(t), \tag{3.11}$$

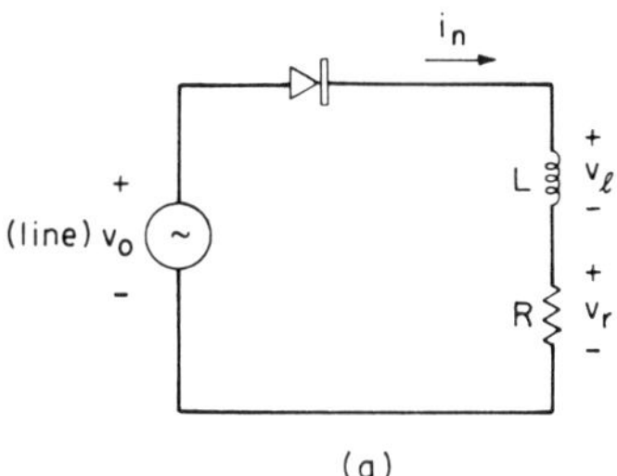

(a)

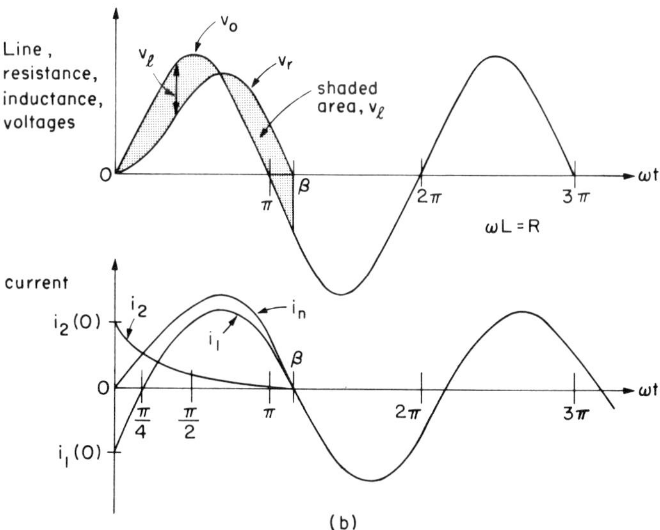

(b)

Figure 3.4   Half-wave circuit with LR load.

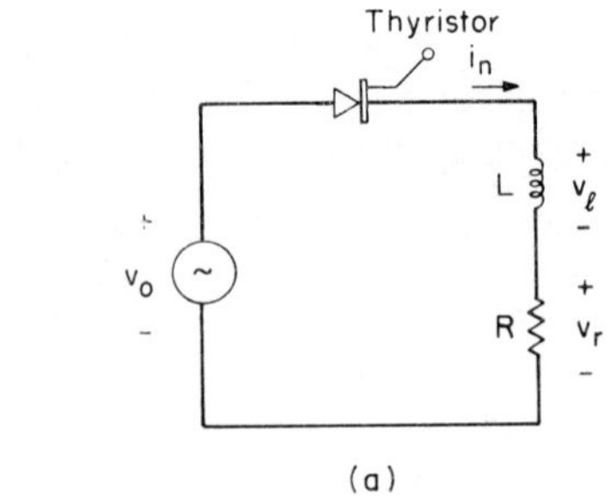

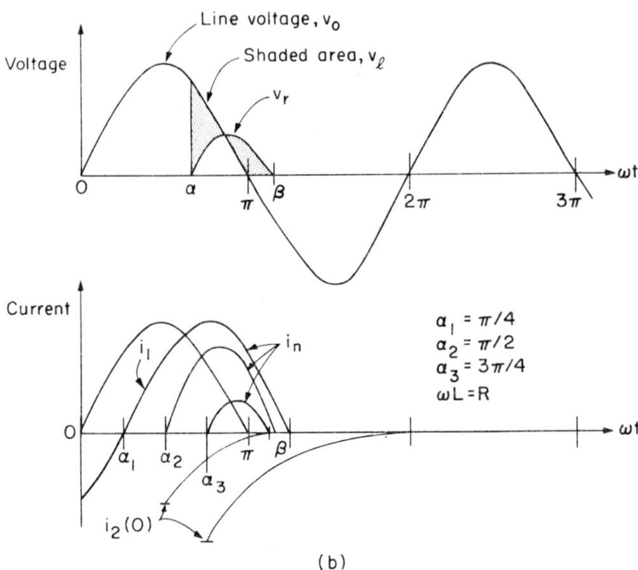

Figure 3.5   Controlled rectifier with LR load.

where

$$i_1(t) = \frac{\sqrt{2}\,V_0}{(R^2 + X^2)^{1/2}} \sin(\omega t - \phi), \qquad (3.12)$$

$$\phi = \arctan \frac{X}{R},$$

and

$$i_2(t) = -i_1(\alpha)e^{-t/\tau}, \qquad \tau = L/R, \qquad (3.13)$$

$$i_1(\alpha) = \frac{\sqrt{2}\,V_0}{(R^2 + X^2)^{1/2}} \sin(\alpha - \phi). \qquad (3.14)$$

The three current waveforms sketched in Figure 3.5 show the two parts of the current for $\alpha = \pi/4$, $\pi/2$, and $3\pi/4$, for $R = \omega L = 1$ pu. The transient current $i_2(t)$ is zero for $\alpha = \pi/4$. The current amplitude and average value is seen to decrease as α is shifted from zero to π. The average current is given by

$$I_n = \frac{1}{2\pi R} \int_\alpha^\beta \sqrt{2}\,V_0 \sin \omega t \; d(\omega t). \qquad (3.15)$$

The expression for the current must be solved for the extinction angle β before the average current can be evaluated.

The circuit for half-wave rectifier with parallel RC load is shown in Figure 3.6. As previously mentioned, the load circuit is the analog of the dc machine armature; the capacitance represents inertia, the resistance represents mechanical damping or load. Current is the analog of torque and voltage of speed.

The diode conducts only when the line voltage v_0 can be more positive than the load voltage v_n and the current i_n in a positive direction. When the diode is conducting, the load voltage is identical with the line voltage. When the diode is blocking, the load voltage is determined by the behavior of the RC elements.

The waveforms in Figure 3.6 are sketched for the arbitrary condition of $R = 1/\omega C$. The time constant τ of the load circuit is given by $\tau = RC = 1/\omega$. The waveforms are meaningful if followed over a one-cycle period of the line voltage.

At $\omega t = 0$, the capacitor C is practically discharged, the load voltage v_n is zero, and the diode starts to conduct when v_0 becomes positive. The current i_n jumps to the value $i_n(0)$ and the capacitor commences charging. The diode current i_n proceeds to supply the capacitor current i_c and the resistor current i_r until the line voltage v_0 reaches its peak at $\omega t = \pi/2$. The diode current is described by the expression

$$i_n(t) = \frac{\sqrt{2}\,V_0}{(R^2 + X^2)^{1/2}} \sin(\omega t - \phi) \qquad (0 < \omega t < \beta) \qquad (3.16)$$

where

$$X = -1/\omega C,$$

$$\phi = \arctan \frac{X}{R}.$$

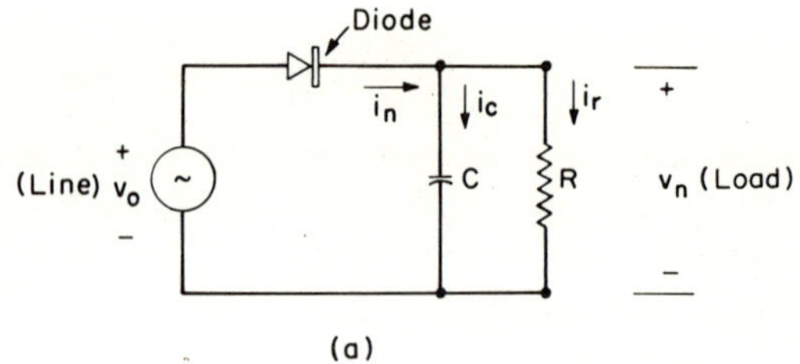

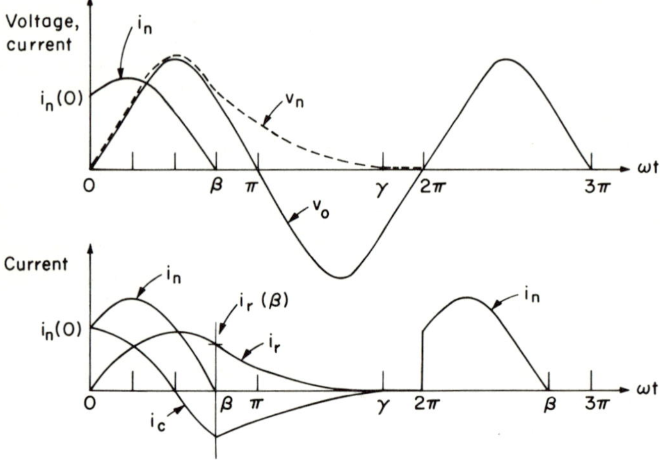

Figure 3.6 Operation of half-wave rectifier with resistance-capacitance load.

From $\omega t = \pi/2$ to $\omega t = \beta$, when $i_n(t) = 0$ and the diode blocks, the capacitor current i_c is negative, meaning that the resistor current i_r is supplied from both the diode current i_n and the capacitor. After the diode blocks, the capacitor supplies all of the resistor current until the capacitor is fully discharged at $\omega t = \gamma$. The resistor current is

$$i_r(t) = -i_c(t) = i_r(\beta)e^{-t/\tau} \qquad (\beta < \omega t < \gamma) \tag{3.17}$$

where

$$i_r(\beta) = \frac{\sqrt{2}\,V_0 \sin \beta}{R},$$

$$\tau = RC.$$

The load voltage has three values over a cycle. They are

$$0 < \omega t < \beta: \qquad v_n = v_0; \tag{3.18}$$

$$\beta < \omega t < \gamma: \qquad v_n = i_r(t)R; \qquad\qquad (3.19)$$

$$\gamma < \omega t < 0: \qquad v_n = 0. \qquad\qquad (3.20)$$

If the time constant of the load is sufficiently large compared to the period of a cycle that the capacitor does not fully discharge by $\omega t = 2\pi$, that is $\gamma > 2\pi$, then the diode starts to conduct when the rising voltage v_0 reaches the declining voltage v_n at $\omega t = (\alpha + 2\pi)$ and the load voltage is never zero. The extreme case occurs for $RC \to \infty$: the capacitor charges to the peak value $\sqrt{2}\, V_0$ and remains there.

In the analog of the load circuit, positive capacitor current indicates that electromagnetic torque is accelerating the armature and load inertia. Negative current designates that the inertia is decelerating and providing torque for the load and the mechanical losses.

3.3   *Three-Phase Rectifier–Resistance Load*

There is an infinite combination of rectifier circuits and load circuits which could be analyzed in detail. However, only two arrangements will be described here, first with resistance load, then with parallel RC load to simulate a dc machine armature. The reader can analyze other circuits or refer to a book like Schaefer[1] for further details.

Half-wave and full-wave bridge circuits will be described for both diode and thyristor operation. The half-wave circuit represents the least expensive way to rectify three-phase voltages, but it requires a neutral wire or a transformer to establish a neutral. The full-wave circuit uses six rectifier elements; the circuit is described as an incomplete bridge when three elements are diodes and three are thyristors.

The three-phase half-wave circuit is shown in Figure 3.7a. The load is supplied by the three elements and returns to the common neutral. Each rectifier element conducts for $120°$ each cycle for diode operation; the load is clamped to the line phase that has maximum voltage during each $120°$ interval.

The waveforms of the line voltages and the load voltage v_n at three firing angles are shown in Figure 3.8. The line voltages are shown with respect to neutral, i.e., v_{an}, v_{bn}, v_{cn}. Only the rectifier element connected to the line at the highest instantaneous voltage can conduct. For diode elements, the diodes assume conduction at the crossover points, $\omega t = \pi/6$, $5\pi/6$, $9\pi/6$, etc. For thyristor elements, the thyristor must be gated to turn on during the interval when its anode-cathode voltage is positive. Otherwise the previously conducting thyristor continues to conduct until its current naturally reaches zero and turns itself off.

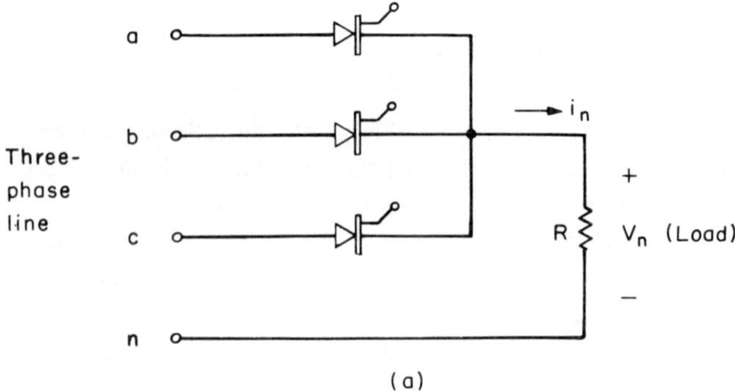

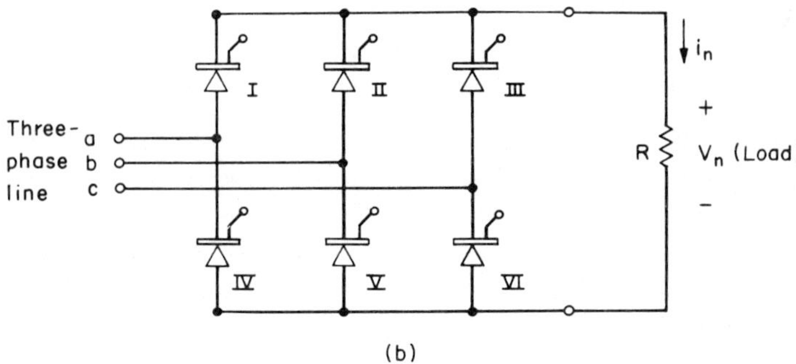

Figure 3.7   Three-phase rectifier circuits: (a) half-wave; (b) full-wave.

The load-voltage waveform for diode operation is shown in Figure 3.8b. The voltage corresponds to the "tops" of the line-to-neutral voltages. The ripple has three pulses per cycle and has a principle third-harmonic component. The average voltage V_n is given by

$$V_n = \frac{3}{2\pi} \int_{\alpha_1}^{\alpha_2} \sqrt{2}\, V_0 \sin(\omega t)\, d(\omega t)$$

$$= \frac{-3}{\sqrt{2}\,\pi} V_0 \cos(\omega t)\Big]_{\alpha_1}^{\alpha_2}. \tag{3.21}$$

For $\alpha_1 = \pi/6,\ \alpha_2 = 5\pi/6$

$$V_n = \frac{3\sqrt{2}}{\pi} V_0\, 0.866 = 1.17 V_0. \tag{3.22}$$

The firing angle for thyristor operation is measured from the point of maximum load voltage; in this case for $\omega t = \pi/6,\ \alpha = 0$. Load current is

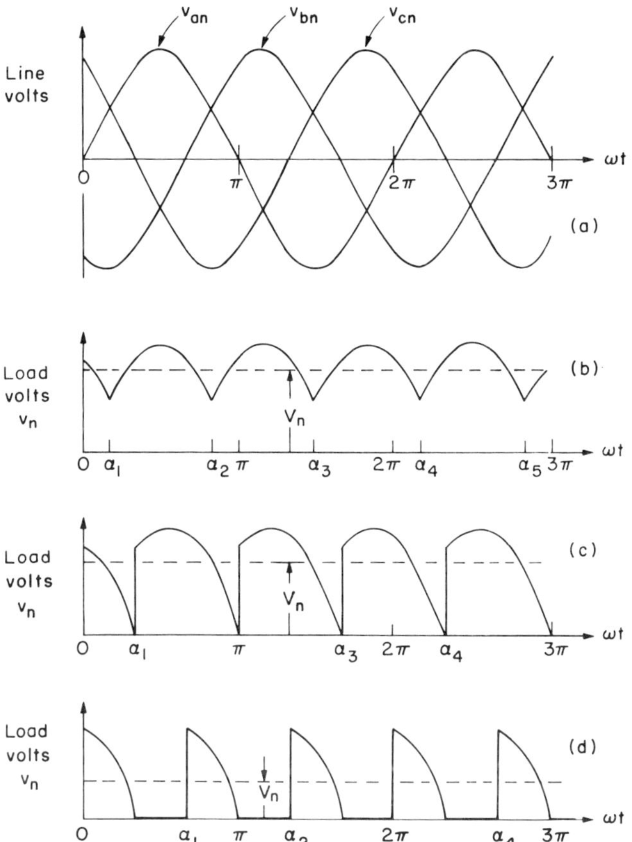

Figure 3.8   Waveforms for three-phase half-wave operation.

continuous for values of α to $\pi/6$, as shown in Figure 3.8c. Beyond $\alpha = \pi/6$, the load current is discontinuous and becomes zero at $\alpha = 5\pi/6$.

For the range $0 < \alpha < \pi/6$, the average voltage is given by

$$V_n = \frac{3}{\sqrt{2}\,\pi}\, V_0[\cos(\alpha + 5\pi/6) - \cos(\alpha + \pi/6)]$$

$$= \frac{3\sqrt{2}}{\pi}\, V_0\, 0.866 \cos \alpha. \tag{3.23}$$

For the range $\pi/6 < \alpha < 5\pi/6$, the average voltage is

$$V_n = \frac{-3}{\sqrt{2}\,\pi} V_0[\cos \pi - \cos(\alpha + \pi/6)]$$

$$= \frac{3\sqrt{2}}{\pi} V_0[0.5 \cos(\alpha + \pi/6) + 0.5]. \qquad (3.24)$$

The plot of the average voltage versus firing angle α is shown in Figure 3.9. The complete range is covered for a change of $5\pi/6 = 150°$.

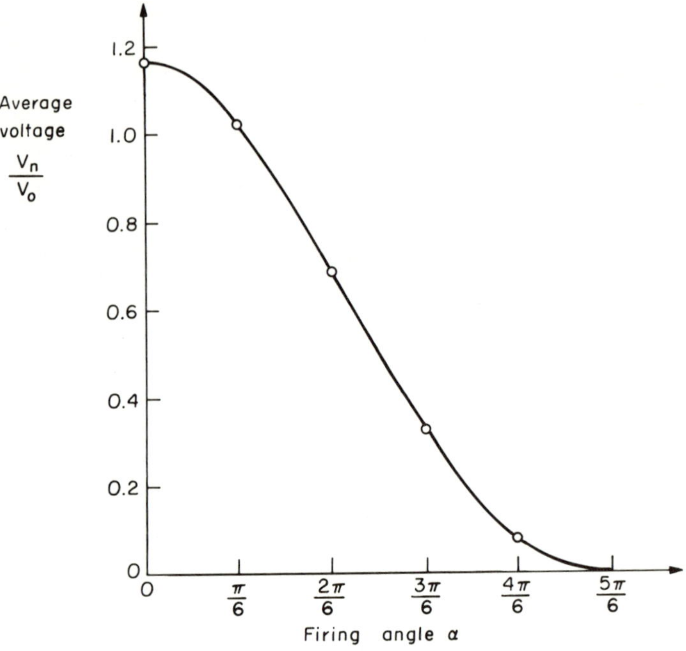

Figure 3.9   Average voltage for three-phase half-wave operation.

The three-phase bridge rectifier circuit shown in Figure 3.7b requires the simultaneous conduction of two rectifier elements to provide a path from the line to the load. When one element of the upper group and one of the lower group conducts, the corresponding segment of the line-to-line voltage is applied directly to the load. Hence, the load voltage is made up of segments of the three-phase line-to-line voltages.

The waveforms of line and load voltages are shown in Figure 3.10 for the three-phase bridge rectifier. Figure 3.10b shows the load voltage for diode rectifier elements or for thyristors fired at $\alpha = 0°$. The load voltage

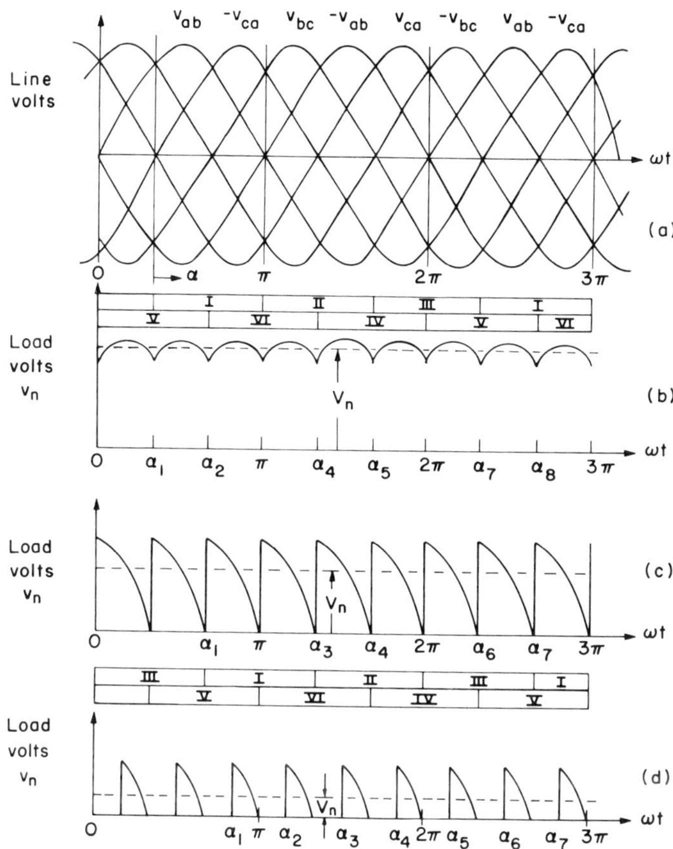

Figure 3.10   Waveforms for three-phase bridge rectifier.

follows 60° portions of the three-phase voltages. From $\omega t = \pi/3$ to $2\pi/3$, diodes I and V are conducting and impressing voltage v_{ab} on the load; from $\omega t = 2\pi/3$ to π, diodes I and VI are conducting impressing voltage $-v_{ca}$ on the load. Each diode conducts for two "pulses" of voltage, or 120°. This represents the maximum voltage condition given by

$$V_n = \frac{-6}{\sqrt{2}\,\pi} \, V_0 \cos(\omega t) \Big]_{\alpha_1}^{\alpha_2}. \tag{3.25}$$

For $\alpha_1 = \pi/3$ and $\alpha_2 = 2\pi/3$

$$V_n = \frac{6}{\sqrt{2}\,\pi} \, V_0 = 1.35 V_0.$$

When thyristors are used in the three-phase bridge circuit, the same requirement for simultaneous conduction of two elements prevails to energize the load. In the range from $\alpha = 0$ to $\alpha = \pi/3$, the load current is continuous and only the on-coming thyristor must be fired at α_1, α_2, etc. The conducting thyristor in the opposite side of the bridge remains conducting. The average voltage is given by

$$V_n = \frac{-6}{\sqrt{2}\pi} V_0[\cos(\alpha + 2\pi/3) - \cos(\alpha + \pi/3)]$$

$$= \frac{6}{\sqrt{2}\pi} V_0 \cos\alpha \qquad (0 < \alpha < \pi/3). \tag{3.26}$$

In the range from $\alpha = \pi/3$ to $2\pi/3$, the load current becomes discontinuous because the off-going thyristor extinguishes itself before the on-coming thyristor is fired. Hence, both thyristors have to be fired to reestablish the current every $60°$. The average voltage is now given by

$$V_n = \frac{-6}{\sqrt{2}\pi} V_0[\cos\pi - \cos(\alpha + \pi/3)]$$

$$= \frac{6}{\sqrt{2}\pi} V_0[1 + \cos(\alpha + \pi/3)] \qquad (\pi/3 < \alpha < 2\pi/3). \tag{3.27}$$

A plot of the average voltage is shown in Figure 3.11. The complete voltage range is covered for a span of α of $120°$.

3.4 *Summary*

Rectifier circuits differ in the phases of the supply, the number of controllable and noncontrollable elements, and the ripple level of the load current. The full-wave bridge circuits are capable of regenerating motor power to the line, although the same can be achieved with two sets of half-wave circuits. The firing circuits determine the start of each conduction period, but the load parameters determine the length of each conduction period.

A comprehensive analysis of all useful rectifier circuits for drives is presented by Schaefer.[1] Bedford and Hoft[2] discuss the rectifier and inverter operation of the three-phase complete bridge.

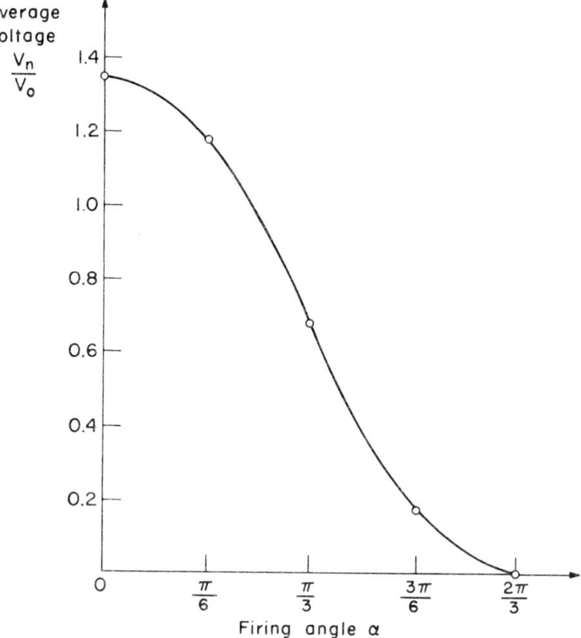

Figure 3.11   Average voltage for three-phase bridge operation.

References

1. J. Schaefer, *Rectifier Circuits, Theory and Design*. New York: John Wiley & Sons, Inc., 1965.
2. B. D. Bedford and R. G. Hoft, *Principles of Inverter Circuits*. New York: John Wiley & Sons, Inc., 1964.

4 SINGLE-PHASE DRIVES

Fractional and low-integral horsepower drives are built to operate from a single-phase line, both because such power is available where such lower-power drives are needed, and because of the lower cost of fewer thyristors. The single-phase drives extend from poorly regulated units with open-loop firing angle control to well-regulated units with armature voltage or tachometer feedback. The series universal motor drives will be treated in Chapter 6.

4.1 *Half-Wave Drives*

The simplest combination of thyristor and motor consists of the half-wave circuit shown in Figure 4.1. The armature is supplied through a

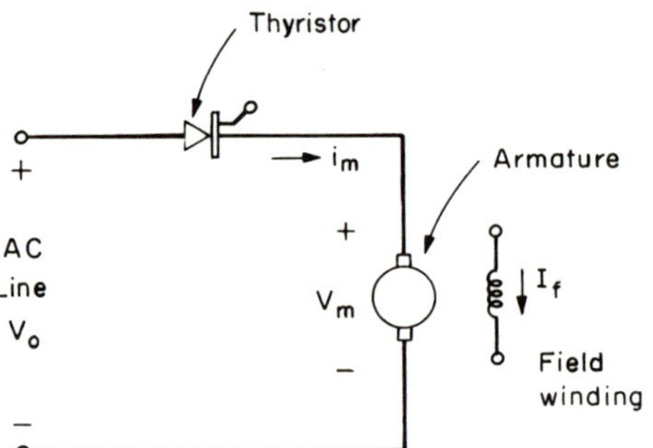

Figure 4.1   Half-wave drive circuit.

single thyristor and may have a freewheeling diode. The field winding is supplied from an independent half-wave or full-wave rectifier. Such a simple circuit combined with a reversing switch and a dynamic braking resistor can provide all of the requirements of a drive system in both directions of rotation.

Waveforms which demonstrate the operating principles of the half-wave drive are shown in Figure 4.2 for the case of a diode instead of the thyristor. The waveforms are for a light-load and a heavy-load condition.

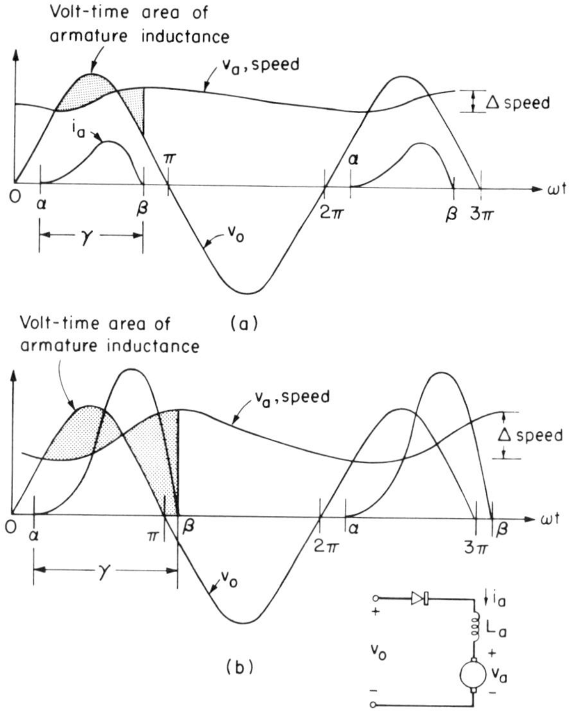

Figure 4.2   Waveforms for half-wave drive with diode:
(a) light load; (b) heavy load.

The waveforms of Figure 4.2 show that the motor operates in the
following manner:

1. The armature current i_a flows in positive pulses between $\omega t = \alpha$ and
   β each cycle for a conduction angle γ.
2. During the time of current flow, the torque produced by the current
   accelerates the motor and load inertia so that the speed rises by Δ
   speed. The lowest speed occurs at α and the highest at β.
3. The motor and load coast down in speed from the end of current
   conduction β in one cycle to the start of conduction α in the next
   cycle. The loss and load torques are provided from the deceleration
   of the inertia during this period

$$T_m = -\frac{\partial}{\partial\theta}\left[\frac{1}{2}J\left(\frac{d\theta}{dt}\right)^2\right] = -J\frac{d^2\theta}{dt^2}. \tag{4.1}$$

The inertia acts as a reservoir of mechanical energy in the half-wave drive.

4. With fixed field current, the armature voltage (emf) is directly proportional to the motor speed.

5. The diode starts to conduct armature current i_a at angle $\omega t = \alpha$ when the instantaneous line voltage v_0 rises to and exceeds the armature voltage v_a, so that the diode becomes forward biased. The diode blocks at angle $\omega t = \beta$ when the line voltage v_0 falls below the motor voltage v_m. The motor voltage is the sum of the armature voltage and the impedance voltage, $v_a + R_a i_a + L_a \, di_a/dt$.

6. The armature and circuit resistance and inductance absorb the difference between the line voltage and the armature voltage in the conduction period. The energy absorbed by the inductance from α to the current peak is delivered to the armature from the current peak to $\omega t = \beta$. As a matter of fact, when the energy is all delivered, the current i_a reaches zero and the diode blocks, ending the conduction period.

7. An increase of motor load, as shown in Figure 4.2b, means that the motor must slow down more in the coasting period and require a larger amplitude current pulse to accelerate during the conduction period. A consequence is that the conduction period γ also increases.

The half-wave operation is characterized by the inductance storing and discharging electrical energy during the diode conduction period, and the inertia storing and discharging mechanical energy between conduction periods. The speed fluctuation shown in Figure 4.2 is exaggerated to show the effect, but it nevertheless must occur to supply the load torque.

The waveforms for the half-wave drive using a thyristor are shown in Figure 4.3. The start of conduction at α requires that the line voltage v_0 equal or exceed the armature voltage v_a, and that the thyristor receive a gating pulse. The operation is identical with that for a diode in regard to the energy storage functions of the inductance and inertia. As the firing angle α is retarded, the speed and voltage v_a must drop to allow the motor to draw the same average current from the line and produce the same load torque. The thyristor reverts to the blocking state when the inductance has discharged its energy and the current falls to zero at $\omega t = \beta$.

The behavior of the circuit can be expressed in terms of average electrical quantities. The equation for the motor circuit during the conduction interval is

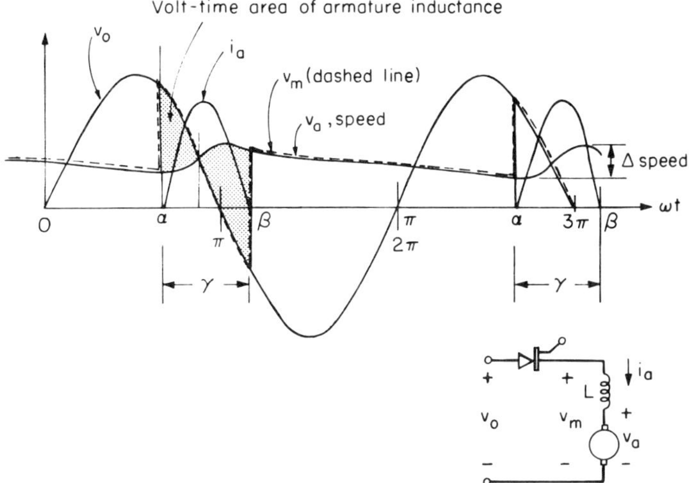

Figure 4.3   Waveform for half-wave thyristor drive.

$$v_0 = i_a R_a + L \frac{di_a}{dt} + v_a. \tag{4.2}$$

Equation 4.2 can be rewritten as

$$v_0 \, dt = R_a i_a \, dt + L \, di_a + v_a \, dt, \tag{4.3}$$

and integrated over the conduction interval

$$\int_{\alpha/\omega}^{\beta/\omega} v_0 \, dt = R_a \int_{\alpha/\omega}^{\beta/\omega} i_a \, dt + L \int_{i_a} di_a + \int_{\alpha/\omega}^{\beta/\omega} v_a \, dt,$$

$$V_m{}' = R_a I_a{}' + V_a' = R_a I_a' + K_a \Phi_f N'. \tag{4.4}$$

The average values in Equation 4.4 are taken only over the conduction interval. They are *not* the average values over a cycle as would be measured with instruments, unless $V_m{}'$ is the measured value on the motor side of the rectifier element.

The equation for the mechanical system is

$$T_m = K_t \Phi_f i_a = T_l + J \frac{d\omega_m}{dt}. \tag{4.5}$$

The equation can be integrated over the cycle to yield

$$K_t \Phi_f \int_0^{2\pi/\omega} i_a \, dt = \int_0^{2\pi/\omega} T_l \, dt + J \int_{\omega_m} d\omega_m,$$

$$K_t \Phi_f I_a = T_l. \tag{4.6}$$

Equation 4.6 shows that the load torque developed by the motor is proportional to the average current I_a over a cycle. The storing operation of the inertia does not appear in the integrated expressions but appears in the instantaneous expressions.

The waveform seen directly across the motor terminals is indicated in Figure 4.3 by v_m. It is the line voltage v_0 during the conduction interval and the armature voltage v_a during the coasting interval.

The disadvantages of the half-wave drive are the following:

1. Because of the short conduction angles, the current has a high rms-to-average ratio so that the motor is limited to less than rated torque at its rated rms current. Some improvement can be obtained with a freewheeling diode across the armature.
2. Under heavy torque load at low speed, the motor will tend to chug because the power is delivered to the motor in pulses once per cycle.
3. A supply transformer will saturate from the dc component in the line current drawn by the half-wave drive circuit.

The advantages are low cost and simplicity.

4.2 Full-Wave Drives

A full-wave drive circuit is usually built as shown in Figure 4.4 using

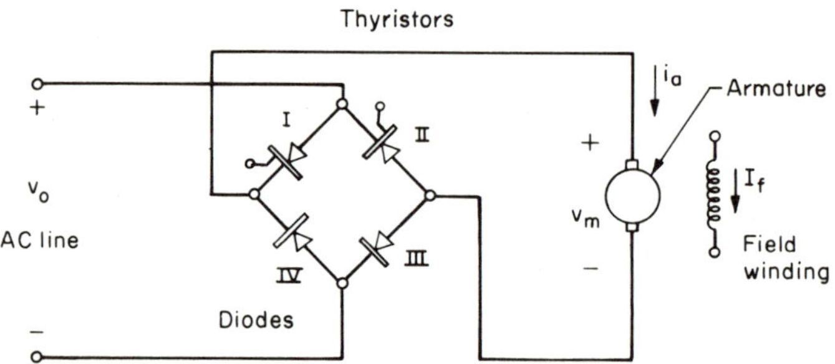

Figure 4.4   Full-wave thyristor drive circuit.

two thyristors and two diodes. The diodes are placed so that they also serve to provide a freewheeling current path across the armature terminals. One thyristor, No. I, is fired on the positive half-cycle of line voltage; current returns to the line through diode No. III. On the negative half-cycle, thyristor No. II is fired and the current returns through diode

No. IV. When the armature attempts to reverse its voltage and follow the line voltage, diodes Nos. III and IV conduct impressing a short-circuit freewheeling path across the armature.

The waveforms of voltage and current for a typical operating point are shown in Figure 4.5. The motor is now supplied with two pulses of power

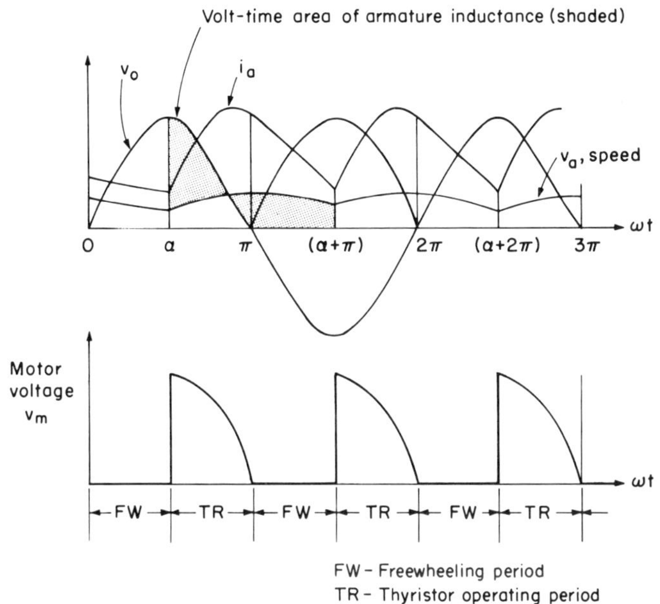

Figure 4.5   Waveforms for full-wave drive.

per cycle and spends less time in the coasting condition. At $\omega t = \alpha$, thyristor No. I is fired; the instantaneous line voltage v_0 is impressed on the armature and causes the current i_a to build up as shown. The resultant torque also accelerates the motor causing its speed to rise by Δ speed. Finally, at $\omega t = \pi$, the armature voltage attempts to reverse, turns on the freewheeling diodes, and the armature current continues to flow either until the volt-time area is balanced, or thyristor No. II is fired.

The disadvantage of the circuit of Figure 4.4 is that the gate terminals of the thyristors cannot be driven with respect to a common cathode. A pulse transformer must be used with two independent secondary windings. An alternative circuit is shown in Figure 4.6 in which the thyristors have a common cathode connection, but a separate freewheeling diode must be used. In small motors, the armature L/R ratio may be so small that a

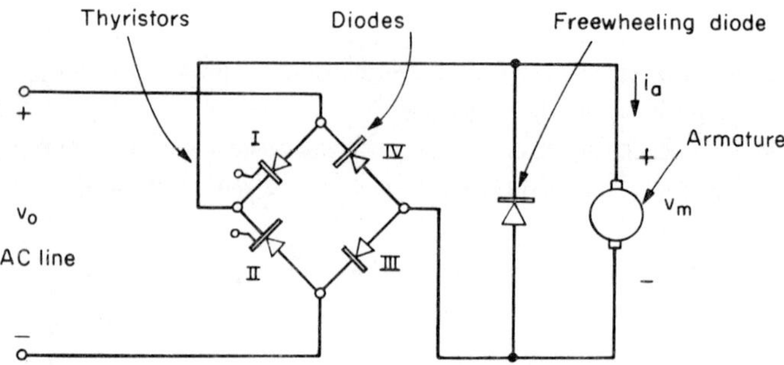

Figure 4.6   Full-wave thyristor drive circuit—common cathode.

freewheeling diode will not contribute any assistance to increasing the conduction angle. However, it may provide some protection to the bridge if the line source is suddenly interrupted.

The term *discontinuous* is applied to the condition where the armature current reaches zero each half-cycle before the next thyristor is fired. Operation in this condition is similar to half-wave operation. The term *continuous* is applied to the condition where the armature current never ceases, but continues to flow either through the thyristors or the diodes. Energy delivered to the armature circuit when the thyristors are conducting is partially stored in the inductance, partially stored in the kinetic energy, and partially used to supply the mechanical load. During the freewheeling period, the energy is recovered from the inductance and is converted to mechanical form to supplement the kinetic energy in supplying the mechanical load. The freewheeling armature current continues to produce electromagnetic torque in the motor.

The armature-circuit equation for the conducting period is

$$v_0 = v_a + R_a i_a + L_a \frac{di_a}{dt}, \qquad \alpha < \omega t < \pi. \tag{4.7}$$

For continuous armature current, the equation for the freewheeling period is

$$0 = v_a + R_a i_a + L_a \frac{di_a}{dt}, \qquad \pi < \omega t < \alpha + \pi. \tag{4.8}$$

We can integrate each of the equations over the respective intervals

$$\int_{\alpha/\omega}^{\pi/\omega} v_0 \, dt = \int_{\alpha/\omega}^{\pi/\omega} v_a \, dt + R_a \int_{\alpha/\omega}^{\pi/\omega} i_a \, dt + \int_{i_a} L_a \, di_a, \tag{4.9}$$

$$0 = \int_{\pi/\omega}^{(\alpha+\pi)/\omega} v_a \, dt + R_a \int_{\pi/\omega}^{(\alpha+\pi)/\omega} i_a \, dt + \int_{i_a} L_a \, di_a. \tag{4.10}$$

The last term in Equation 4.9 is the increase of flux linkage over the interval; the corresponding term in Equation 4.10 is the negative value. We can add the two equations together, divide by the time of a half-period, and get

$$V_m = V_a + R_a I_a, \tag{4.11}$$

where

$V_m = 0.45 V_0 (1 + \cos \alpha) = $ average voltage applied to motor;
$V_a = K_a \Phi_f N = $ average armature generated voltage;
$I_a = T/K_t \Phi_f = $ average armature current.

Equation 4.11 can be rearranged to show the speed-torque characteristic of the motor

$$N = \frac{0.45 V_0 (1 + \cos \alpha) - (R_a/K_t \Phi_f)T}{K_a \Phi_f}. \tag{4.12}$$

Equation 4.12 has the familiar form for motor operation from continuous dc sources. The first term is the no-load speed, while the second term is the speed droop produced by the torque T. The no-load speed is varied from maximum value at $\alpha = 0$ to zero at $\alpha = \pi$. However, the equation is valid *only* in the operating regions where the armature current is *continuous*.

Continuous conduction is obviously a desirable condition to obtain best speed regulation and highest average-to-rms current ratio. It is promoted by insuring a freewheeling current path and using an external armature circuit choke to increase the time constant.

4.3   *Gaudet Drive Circuit*

The Gaudet circuit utilizes a single thyristor to present the equivalent of full-wave control to the armature of the dc motor.[1] The circuit is shown in Figure 4.7. It consists of a diode bridge which develops a full-wave rectified voltage v_r, and a single thyristor which is fired at twice the line frequency to control each half-cycle of the rectified voltage. The armature is shunted by a freewheeling diode. The field winding can be supplied from the common full-wave diode bridge.

A set of typical waveforms for the Gaudet circuit is shown in Figure 4.8. The thyristor is fired at $\omega t = \alpha$, $\pi + \alpha$, etc. The thyristor conducts to

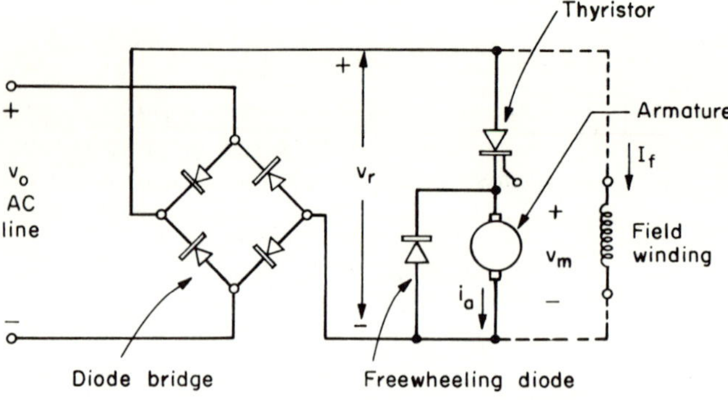

Figure 4.7   Gaudet drive circuit.

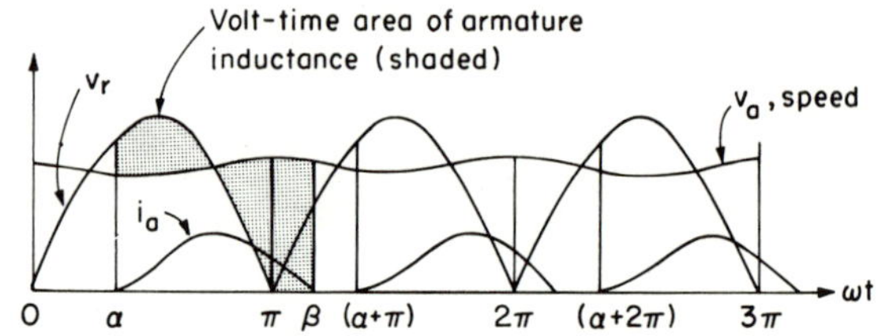

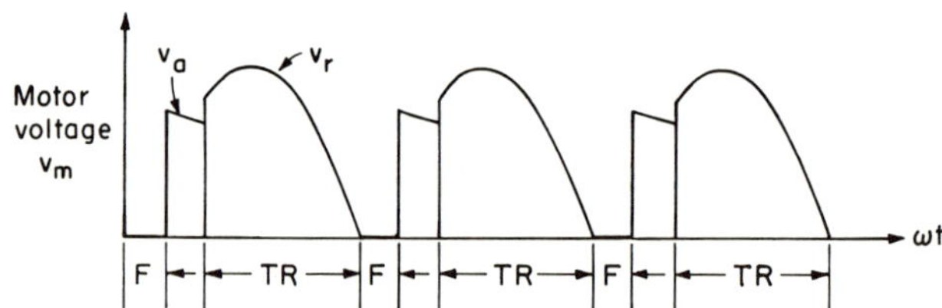

Figure 4.8   Waveforms for Gaudet drive circuit.

$\omega t = \pi$, then blocks, while the current i_a continues through the freewheeling diode to $\omega t = \beta$ until the volt-time areas are equalized. The
current remains zero until the thyristor is fired at $\omega t = \pi + \alpha$ to start the
next half-cycle. For low speeds and/or heavy torques, the conduction can
become continuous. The circuit operates down to zero speed at which
level the drop through the freewheeling diode provides the potential to
block the thyristor. The feature of the circuit is the economy of one
thyristor to obtain full-wave operation. Also, the same bridge rectifier can
serve the armature and field circuits.

4.4   *Discontinuous Conduction*

As shown in Section 4.2, the armature current for a full-wave rectifier can flow in discrete pulses or continuously with a superimposed ripple. The term discontinuous is applied to the first case and continuous to the latter. Discontinuous conduction results in a high form factor but more importantly causes the open-loop, speed-torque curves of the motor to fall in two regions. The speed falls rapidly with load torque when the current is discontinuous. In a regulating system, considerably more loop gain is required than in the region of continuous conduction.

The speed-torque curves in the region of continuous conduction can be calculated merely from the average values of voltage and current. However, in the discontinuous region, the calculations are cumbersome. For each value of α and speed, the expression for the instantaneous armature current must be found from $t = \alpha/\omega$ to β/ω, and then integrated to find the average value and the torque. Pfaff[2] shows the procedure and results of a computer calculation.

The nature of the phenomena can be determined without undertaking the calculations by using graphical methods and a simple model of the motor, as shown in Figure 4.9. Assume that the supply is a controlled full-wave single-phase rectifier with a freewheeling diode, and that the motor armature has only inductance, no resistance. The speed-torque curve for a given value of α can be estimated quickly.

Consider the firing angle $\alpha = 90°$; select an armature voltage v_a corresponding to a speed N_1. The angle of conduction is determined by equating the shaded volt-time areas between v_0 and v_a corresponding to the inductance voltage. The shaded volt-time area is proportional to the amplitude of the armature current I_1. The area under the current is proportional to the average current and the torque T_1. Mark the point $N_1 - T_1$ on the speed-torque curve.

As the selected speed is lower and lower, a speed such as N_4 is reached where the conduction angle of the current is $180°$. For that speed, and for any heavier loadings, the current is continuous. In our model, the speed will be constant, but in an actual motor the speed will drop slightly because of the armature resistance.

The construction process can be repeated for other firing angles, for example, $\alpha = 45°$ and $135°$. The three cases are summarized in Figure 4.9: as the firing angle approaches $180°$, the speed drop to the continuous conduction region becomes steeper. If the rectifier is fired early enough, and the motor has sufficient load, the drive will operate only in continuous conduction.

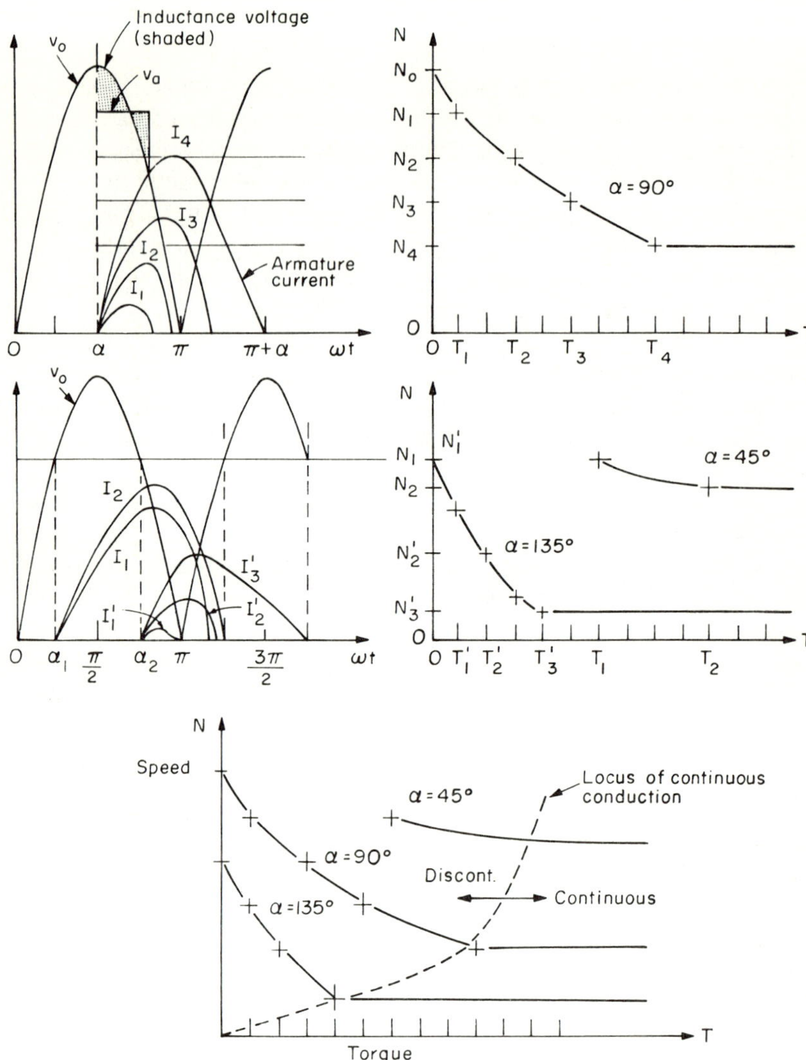

Figure 4.9   Construction of speed-torque curves for single-phase
full-wave drive circuit with freewheeling diode.

Not only does discontinuous conduction yield a high form factor and
excessive speed regulation, but Black[3] shows that the speed dip on impact
loading is severe when the drive falls into and must recover from a
discontinuous conduction region. He derives expressions for the induct-

ance to be added to the armature circuit of motors to insure continuous conduction at low currents, and improve the speed response to impact loads.

4.5 *Commercial Aspects*

Single-phase commercial drives are supplied in either half-wave or full-wave circuits, and for 115-V or 230-V operation. The lowest horsepower, least expensive units are designed for 115-V, half-wave operation. The highest horsepower in the range are designed for 230-V, full-wave operation. Maximum allowable supply line current limits the horsepower.

The 115-V half-wave drives are built up to about one horsepower; the full-wave drives are built to about five horsepower. Within the range fractional horsepower sizes of 1/8, 1/6, 1/4, 1/3, 1/2, and 3/4 are built. The standard base motor speeds are 1150, 1750, 2500, and 3500 rpm.

The control systems provide speed ranges of up to 20:1 for rated motor torque, and up to 100:1 for reduced motor torque. The standard system uses armature voltage as a measure of speed and attains speed regulation at base speed of up to 3 per cent. With tachometer speed sensing, regulation as fine as 0.1 per cent of base speed can be obtained.

The 230-V full-wave drives typically cover 1, 1½, 2, 3, and 5 horsepower. The performance is of the same order as for the 115-V drives but depends to a considerable extent on the manner in which the various manufacturers rate their equipment.

In addition to the basic drive, optional equipment can be provided for extending the speed range by field weakening, acceleration limit, braking, reversing and speed setting. The drives can also be supplied with various types of motors such as drip-proof and blower ventilated.

Commercial drives employ basically the same ratings of solid-state devices for the same motor ratings. They differ on the manner of generating the firing signals, protecting the SCR's and in packaging arrangement. There is considerable difference in ease of repair and maintenance from one manufacturer to another. A simplified functional schematic diagram of a single-phase system is shown in Figure 4.10

4.6 *Summary*

Single-phase drives are operated half-wave or full-wave and usually with discontinuous armature current. The smaller, fractional-horsepower drives are frequently operated without speed feedback; the larger ones use armature-voltage sensing for speed control.

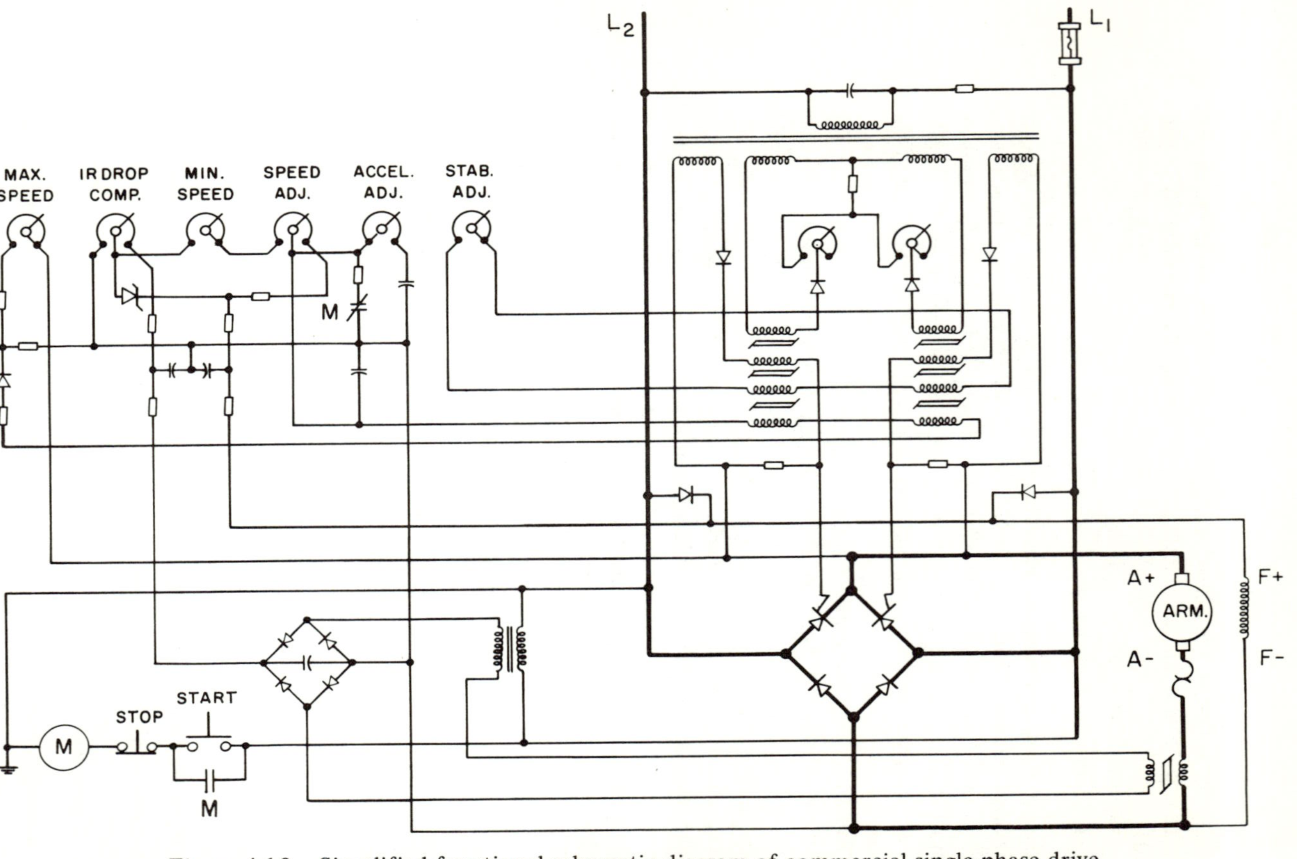

Figure 4.10 Simplified functional schematic diagram of commercial single-phase drive circuit. (Courtesy of Electric Regulator Corp., Norwalk Conn.)

56

The operating principles and typical commercial information on single-phase drives is given by Kusko.[4] A single-phase reversing drive using bilateral thyristors has been described by McMurray;[5] his paper includes pertinent references as well.

References

1. J. A. Gaudet, "Motor speed control system," U.S. Patent No. 3,123,757, Issued March 3, 1964.
2. R. W. Pfaff, "Characteristics of phase-controlled bridge rectifiers with d-c shunt motor load," *AIEE Trans. Appl. Ind.,* vol. 77, pt.II, pp. 49–53, May 1958.
3. K. G. Black, "The effect of rectifier discontinuous current on motor performance," *IEEE Trans. Ind. Gen. Appl., pp. 377–382, November 1964.*
4. A. Kusko, "Packaged SCR dc drives," *Control Eng.* June, July 1966.
5. W. McMurray, "Reversing dc motor drives using bilateral thyristors," *Conference Record of the 1966 First Annual Meeting of the IEEE Industry and General Applications Group,* pp. 557–566.

5 THREE-PHASE DRIVES

The bulk of integral-horsepower solid-state dc drive systems utilize three-phase thyristor rectifier circuits for providing the motor power. These circuits are primarily incomplete bridges; complete bridges are used for regeneration; half-wave circuits are used for economy and reversing. These circuits have been described for static loads in Chapter 3.

5.1 *Incomplete Bridge*

The three-phase incomplete bridge circuit is most commonly used for dc motor drives because it requires only three thyristors to achieve the full range of control. The circuit is shown in Figure 5.1. The remaining three

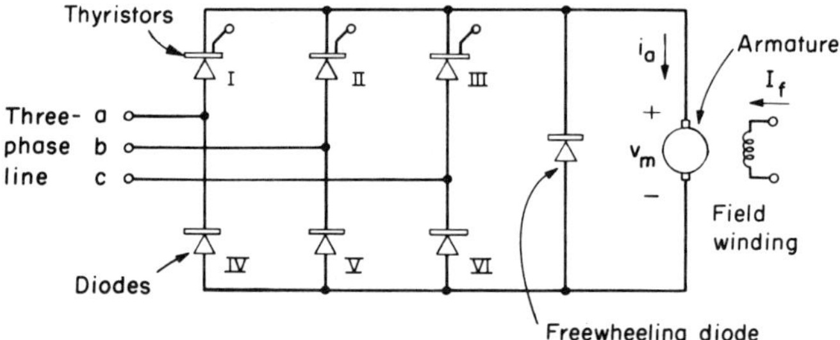

Figure 5.1   Three-phase incomplete bridge circuit.

elements of the bridge are diodes. The complete bridge rectifier is used where the motor is reversed and regenerated to the line. It also has less ripple at small conduction angles. The incomplete bridge is usually operated with a freewheeling diode to increase the conduction angle of the current for low-speed, high-torque operation.

The voltage applied to the armature is controlled by the firing angles of the thyristors I, II, and III in Figure 5.1. The diodes IV, V, and VI merely provide a return path for the current to the most negative line terminal. The difference between the instantaneous voltage V_m applied by the bridge to the motor and the armature voltage V_a is absorbed by the armature circuit inductance primarily, and by the resistance secondarily. The freewheeling diode operates at the end of a current conduction period when the armature voltage attempts to reverse as the inductance discharges its energy.

The operation of the circuit is shown in a sequence of waveforms for various firing angles in Figure 5.2. For a firing angle of $\alpha = 0$, the motor voltage would appear as the usual six-pulse per cycle rectifier output, as shown in Figure 3.10b. As the firing angle is retarded to $\alpha_1 = 30°$, as shown in Figure 5.2b, the firing of the thyristors is delayed, but not of the return diodes, so that only alternate pulses are affected. The armature

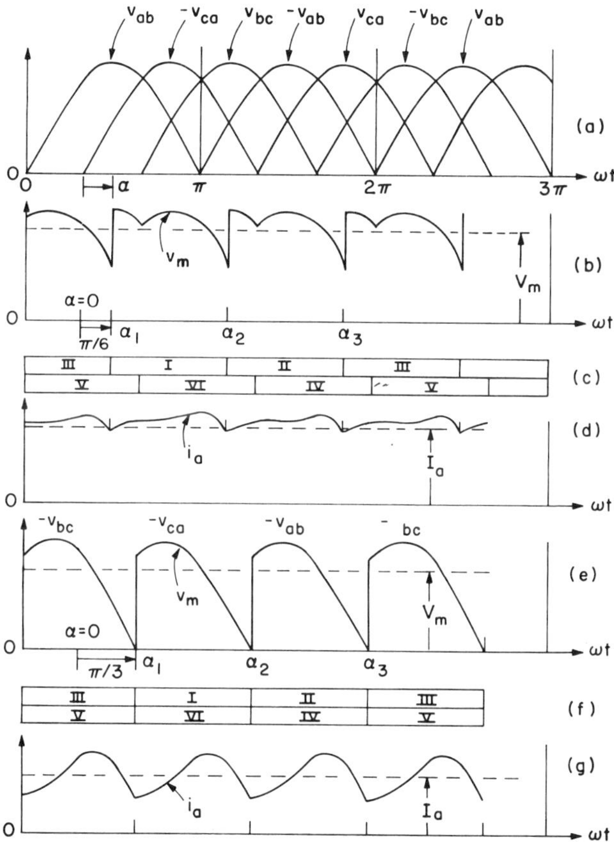

Figure 5.2   Waveforms for incomplete bridge operation.

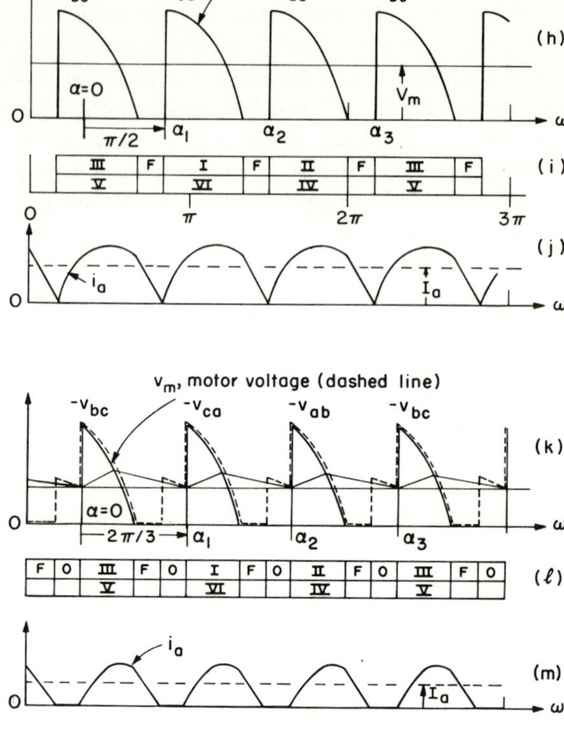

Figure 5.2   *Continued.*

current is continuous and has little ripple. The average armature current I_a is given by

$$I_a = \frac{V_m(\alpha) - V_a(N_1\Phi_f)}{R_a}.$$ (5.1)

So long as the conduction is continuous, the average quantities obey Ohm's law.

For the firing angle shifted to $\alpha_1 = 60°$, as shown in Figure 5.2e, the thyristors are fired so that the current only returns through one diode during each $120°$ conduction interval. The waveform of motor voltage is identical with that of the three-phase half-wave bridge, Figure 3.8c, for $\alpha = 30°$. As a matter of fact for all firing angles greater than $\alpha_1 = 60°$, the waveforms for the two arrangements are identical except that the incomplete bridge uses portions of line-to-line voltage whereas the half-wave circuit uses portions of the line-to-neutral voltages.

For the firing angles retarded to $\alpha_1 = 90°$, as shown in Figure 5.2h, the conduction period through the bridge elements is less than $120°$ per pulse. The current completes its conduction period and discharges the inductance by forcing the freewheeling diode to conduct as shown in the sequence, Figure 5.2i. Without the freewheeling diode, the motor voltage would go negative, the current would reach zero more quickly and be discontinuous, and the energy stored in the inductance would be partially returned to the supply line and the remainder delivered to the armature.

The last firing angle, shown in Figure 5.2k is for $\alpha_1 = 120°$. The motor voltage each pulse has three components. From $\omega t = 120°$ to $180°$, the thyristor conducts and applies a section of line voltage to the motor. For a period thereafter, the freewheeling diode conducts and holds the motor voltage at zero. When the inductance is all discharged, the freewheeling diode blocks and the motor voltage rises to its generated armature voltage value. As shown in Figure 5.2m, the armature current is discontinuous.

The incomplete bridge requires a range of firing angle from $\alpha_1 = 0$ to $\alpha_1 = 180°$ to obtain full control. The waveform up to about $\alpha_1 = 30°$ is predominantly six pulse per cycle. From $\alpha_1 = 30°$ to $180°$, it is three pulse per cycle. The firing angles at which the current becomes discontinuous as well as the ripple amplitude depends upon the L/R time constant of the armature circuit.

5.2   *Complete Bridge*

The complete bridge circuit is shown in Figure 3.7. It uses six thyristors and no diodes except for freewheeling operation. The thyristors are fired in sequence every $60°$ and the ripple is always six pulses per cycle. The complete bridge produces less ripple than the incomplete bridge, but is primarily used where regeneration is required. If the polarity of the motor armature voltage is reversed with a reversing contactor or by reversing the field current, then the complete bridge can be operated as an inverter to transfer power from the motor to the line and finally to accelerate the motor to its required reverse speed.

The operation of the complete bridge with a motor load and a freewheeling diode produces waveforms just like those for the resistance load shown in Figure 3.10. For firing angles $\alpha_1 > 60°$, the rectifier voltage becomes discontinuous and will be zero between pulses if the armature time constant keeps the current continuous through the freewheeling diode. If not, the armature voltage v_m will jump up to the generated voltage v_a when the freewheeling diode blocks, as shown in Figure 5.2k. The full range of control is obtained with a firing angle range of $120°$.

When the armature circuit has sufficient inductance to maintain continuous armature current and no freewheeling diode is used, then the armature voltage will go negative for $\alpha_1 > 60°$ to balance the volt-time area of the inductance. The extreme case is shown in Figure 5.3b for $\alpha_1 = 90°$. The average motor voltage V_m is now zero and the average current I_a is also zero. Compare this with the waveform of Figure 3.10d for resistance load. The rectifier control range is now $\alpha_1 =$ zero to $90°$ for full control.

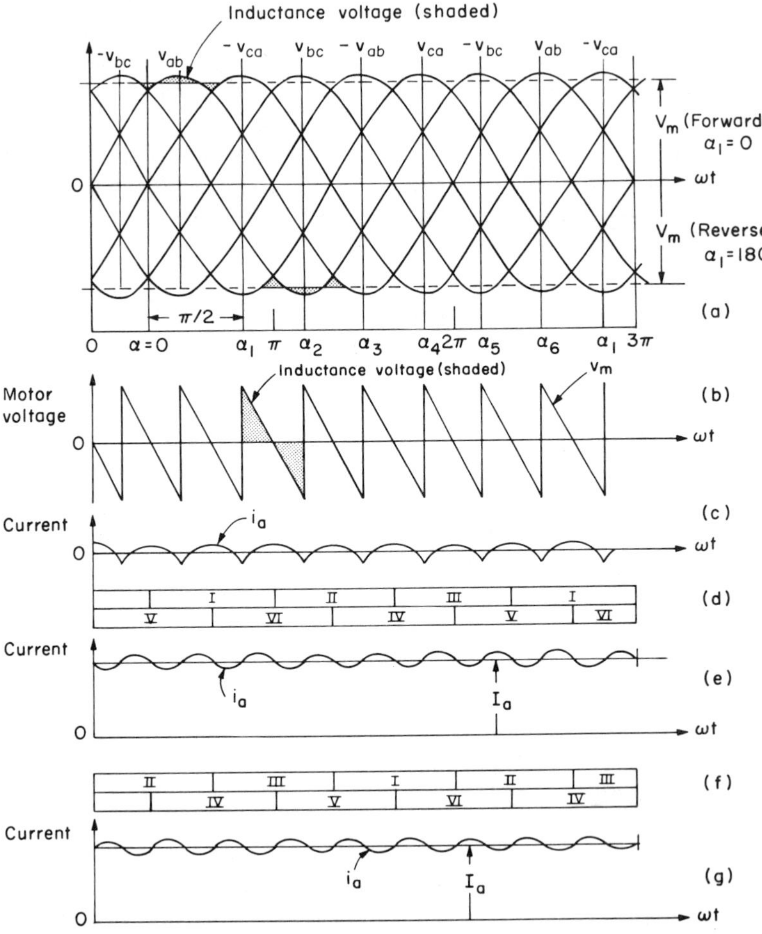

Figure 5.3   Waveforms for complete bridge operation.

If the firing angle is retarded in the range $\alpha_1 = 90°$ to $180°$, the terminal voltage of the bridge will reverse and it will be capable of receiving power from a dc source of the same polarity. The extreme cases of rectifier operation at $\alpha_1 = 0°$ and inverter operation at $\alpha_1 = 180°$ are shown in Figure 5.3a; the currents are shown in Figure 5.3e and g. The direction of the current remains fixed but the voltages reverse.

The directions of the electrical quantities during regeneration are shown in Figure 5.4 for the interval $\omega t = 240°$ to $300°$. The armature is delivering power $v_m i_a$ to the bridge and thence to the line terminals a-b. The voltage–firing angle characteristic is shown in Figure 5.5. As the

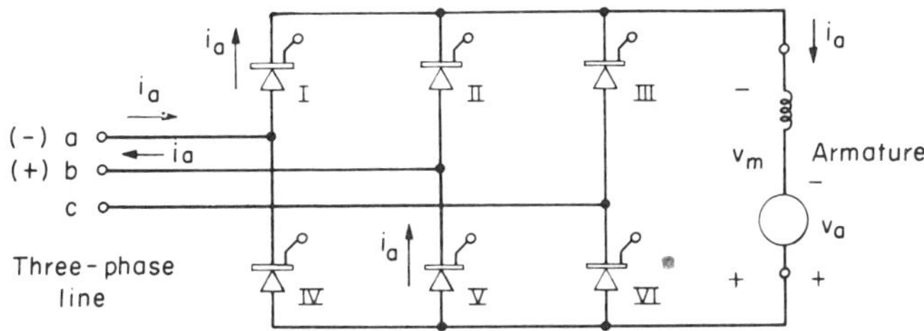

Figure 5.4   Three-phase complete bridge.

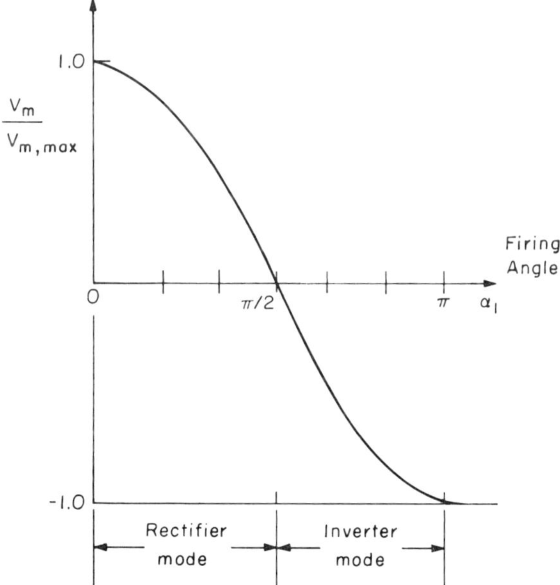

Figure 5.5   Voltage-firing angle characteristic of complete bridge, inductive load.

motor slows down, the firing angle must be advanced to keep the current up and regenerate the power. The motor must be reversed with means other than the bridge.

5.3 Dynamics

Two sets of three-phase half-wave circuits connected between a three-phase line and a dc motor are used to obtain reversing operation and regenerative braking without requiring the use of a reversing contactor for field or armature circuit. One set of three thyristors is used to apply forward, positive current; the other set is used to provide reverse, negative current. One or the other set of thyristors is fired at one time, not both.

The simplest form of deceleration for any drive circuit is that of allowing the motor to coast down to its new speed by its own losses or the drag of the load. The waveforms for such deceleration are shown for the half-wave circuit in Figure 5.6. The motor is initially operating at full speed with a firing angle of $\alpha = 60°$. The firing angle is suddenly retarded to $\alpha = 120°$ and the first new pulse α_2 is applied to thyristor II on voltage v_{bn}. The armature voltage v_a is larger than the line voltage so that the thyristor does not fire, but the motor continues to decelerate in the absence of armature current. The next pulse α_3 is applied to thyristor III on voltage v_{cn}. Finally pulses α_4 and α_5 on voltages v_{bn} and v_{cn} produce armature current and the motor stabilizes at a new lower speed.

The deceleration by coasting is uncontrolled and not particularly desirable in larger horsepower drives. The time for deceleration depends upon the mechanical time constant, that is, the inertia and mechanical torque requirements of motor and load. The deceleration can be increased by connecting a braking resistor across the armature terminals to dissipate some of the kinetic energy. The operation of the contactor for the resistor must be integrated with the control system.

A more desirable deceleration procedure is to use regenerative braking, or the orderly extraction of the kinetic energy and its transfer back to the line. The use of the complete bridge for regeneration once the armature voltage has been reversed has been discussed. However, the two sets of half-wave circuits are more suitable.

The deceleration to standstill and acceleration to a new speed in the opposite direction are shown in Figure 5.7. The motor is assumed to be operating initially at full speed in the reverse direction when the firing signals to the reverse circuit are blocked and the signals to the forward circuit are initiated. The forward circuit applies current in a direction to first extract the armature energy, then apply energy to accelerate the motor in the forward direction.

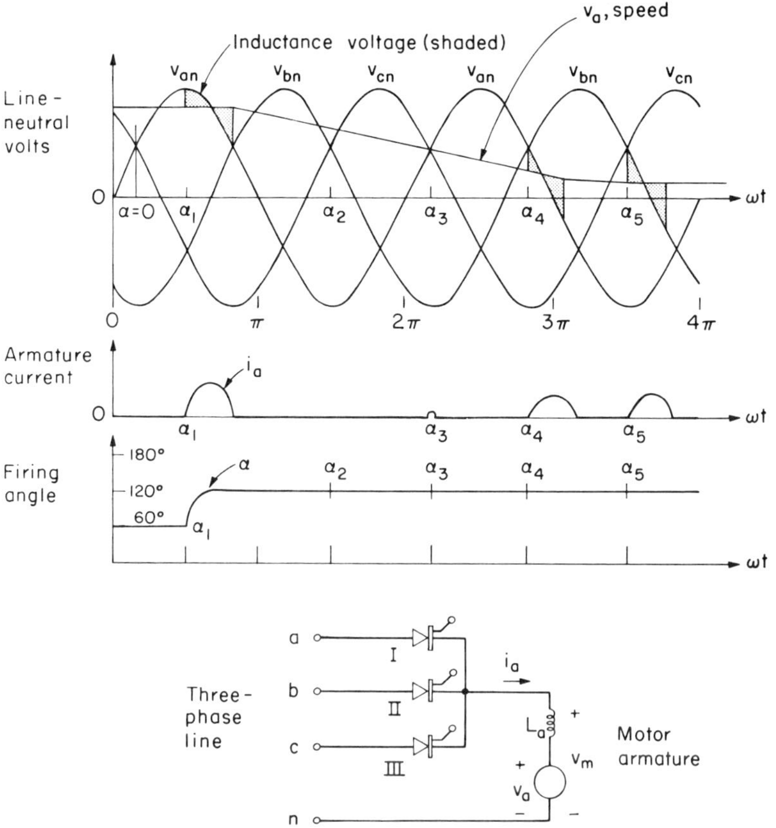

Figure 5.6   Three-phase half-wave circuit under coasting deceleration.

The first pulse is applied to thyristor III on voltage v_{cn} at $\alpha_1 = 180°$, in order to keep the armature current pulse i_a within safe limits. As the motor decelerates, the firing angle is advanced to keep the armature current pulses of the same size. This advancing is controlled automatically by the current-limit circuit. When the firing angle reaches $\alpha_7 = 60°$, the motor has reached its desired forward speed.

If the current pulses and armature voltages are examined pulse by pulse, one will see that the power is negative until the motor reaches standstill at about $\omega t = 420°$, then it becomes positive as the motor receives accelerating power. The firing circuits need a range of at least $180°$ to keep the armature current within limits during deceleration.

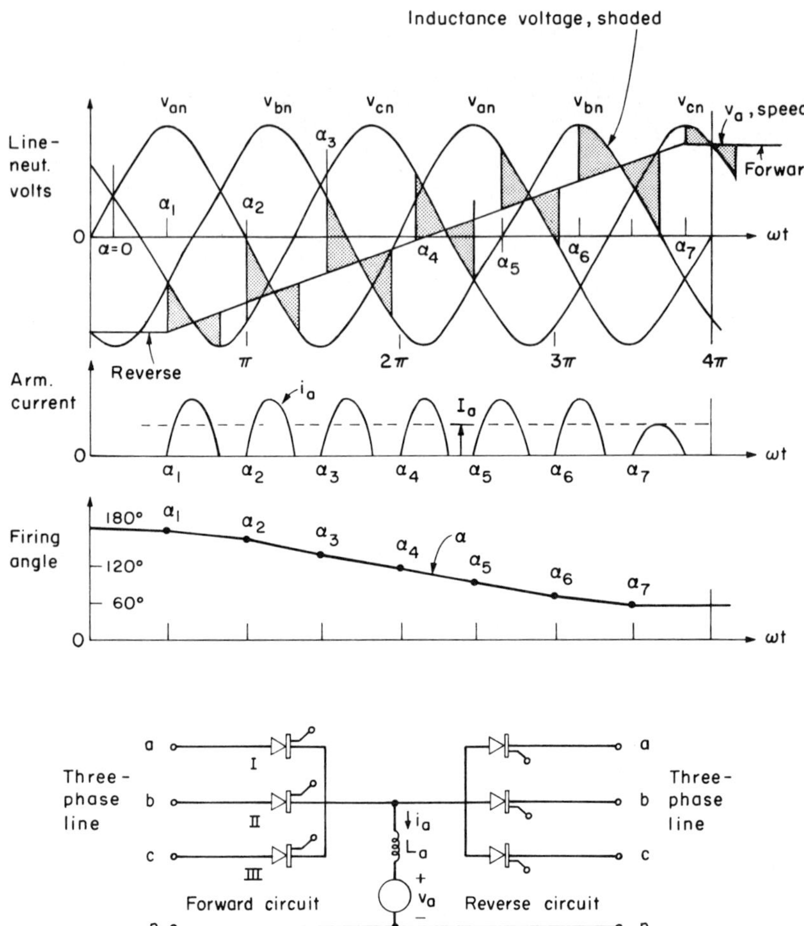

Figure 5.7   Three-phase half-wave reversing circuit.

The same type of firing angle behavior and current pulses will be required for a single-ended drive circuit which is accelerating under current-limit control. The average current I_a during the accelerating period will be held constant by the current-limit circuit.

5.4 *Commercial Equipment*

Commercial three-phase drives are built to operate from either 230-V or 460-V three-phase lines, usually without transformers. The drives utilize complete or incomplete bridges, or half-wave circuits depending upon the

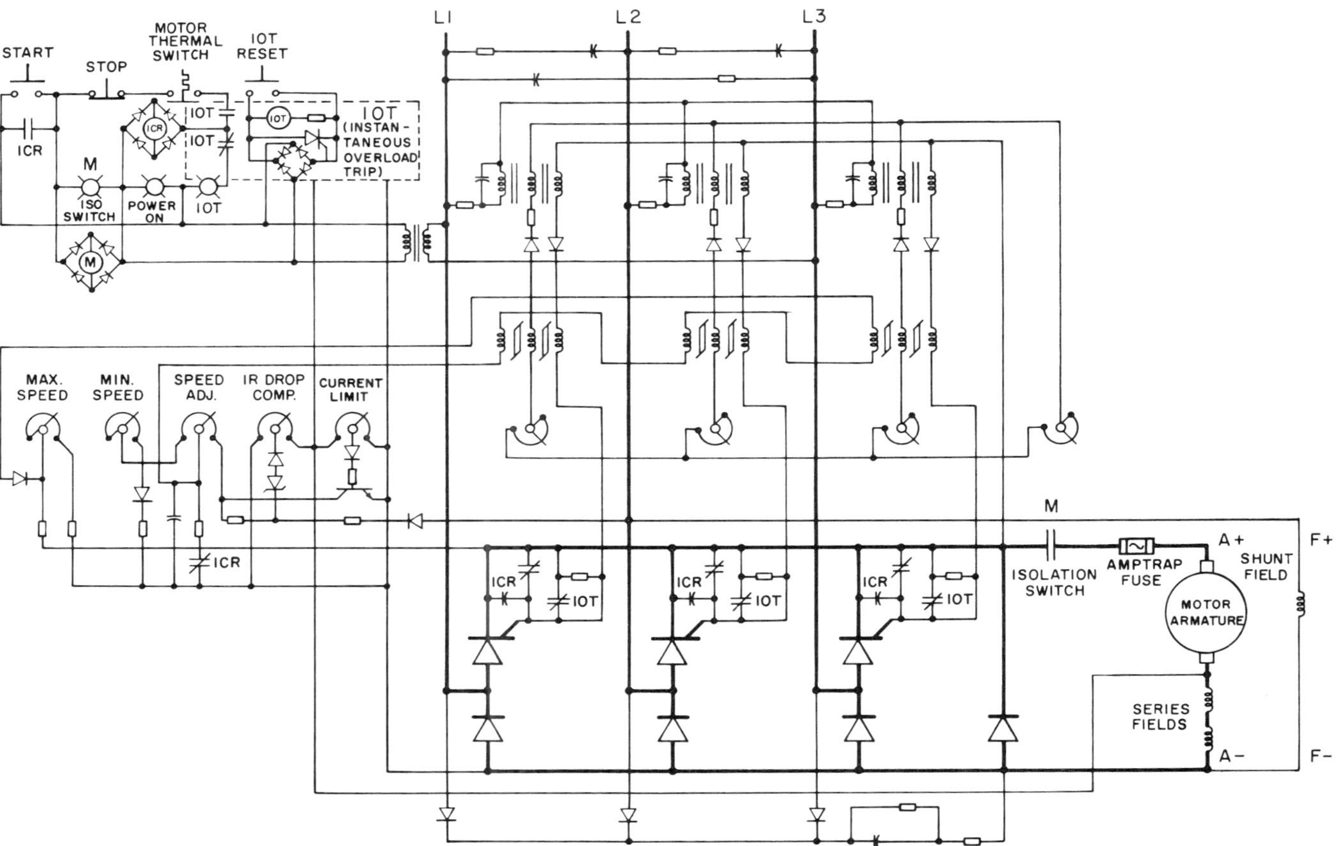

Figure 5.8   Simplified functional schematic diagram of commercial three-phase drive circuit. (Courtesy of Electric Regulator Corp., Norwalk, Conn.)

reversing and braking requirement. The half-wave drives require a neutral or an external grounding transformer to provide the star point.

The range of packaged (nonspecial) drives extends from one to about 150 hp, depending upon the manufacturer. Base motor speeds are typically 850, 1150, 1750, 2500, and 3500 rpm. The larger ratings are obtained by using two or more thyristor bridges in parallel to provide the power.

The typical full-load speed range is 8 to 1, and for light load, 100 to 1. The standard drives utilize armature voltage with IR compensation for speed sensing. Typical regulation is ± 2 per cent of base speed. Speed regulation to ± 0.5 per cent to ± 0.1 per cent is obtained with tachometers instead of armature voltage sensing. Current-limit control is standard, while reversing and speed extension by field weakening are optional features. A simplified functional schematic diagram of a commercial three-phase drive is shown in Figure 5.8.

5.5 Summary

Three-phase thyristor dc motor drives are usually built without transformers by matching the dc motor voltage rating to the rectified value of the supply voltage. Double half-wave circuits are used for reversing drives and incomplete bridges for single direction drives. Freewheeling diodes and armature circuit chokes serve to reduce the ripple factor and armature heating.

The operation and design of a three-phase half-wave reversing drive has been described by Berman.[1] The design of the control system and firing circuits for a drive using a three-phase incomplete bridge has been presented by Wise and Schlieman.[2] Problems in the application, operation, and maintenance are treated by Schofield et al.,[3] and Mellgren.[4] Commercial information is given by Kusko.[5]

References

1. B. Berman, "Static-high capacity-regenerative d.c. amplifier motor control," *Proceedings of the 1965 Intermag Conference,* pp. 16-2-1 to 16-2-7.
2. A. V. Wise and R. G. Schlieman, "A dc motor drive using the new pressure assembled thyristors," *Conference Record of the 1967 Second Annual Meeting of the IEEE Industry and General Applications Group,* pp. 71–79.

3. J. R. G. Schofield, G. A. Smith, and M. G. Whitmore, "The application of thyristors to the control of DC machines," *Proceedings of the 1965 IEE Conference on Power Applications of Controllable Semiconductor Devices,* pp. 219–225.
4. G. Mellgren, "Thyristor converters for motor drives—some experience in design and operation," *Ibid.,* pp. 230–249.
5. A. Kusko, "Packaged SCR dc drives," *Control Eng.* June, July, 1966.

6 SERIES UNIVERSAL MOTOR DRIVES

Series universal motors are used exclusively in portable tools and appliances, where their high rotational speed provides high horsepower per pound. The solid-state speed control systems are simple and economical; they are usually manually controlled to provide the user with a modest speed control facility.

6.1 *Basic Circuits*

Series motor drive circuits are divided into half-wave and full-wave circuits, and also into nonfeedback and feedback circuits. The series motor will produce positive torque for either direction of current so that the motor will respond to either half- or full-wave current.

The elementary half-wave circuit is shown in Figure 6.1. The circuit uses a thyristor for phase control and a simple *RC* firing circuit. The current flows in one pulse per cycle and produces torque only during the pulse. The speed regulation is that of the series motor itself or poorer.

The elementary full-wave circuit is shown in Figure 6.2. The circuit uses a bidirectional ac switch and provides two current and torque pulses per cycle. A simple nonfeedback *RC* firing circuit is shown. For full conduction, the motor operates substantially as though it is connected directly to the ac line. Retarding the firing angle reduces the speed.

The simplest feedback-type half-wave circuit is known as the Momberg circuit and is shown in Figure 6.3.[1] The potentiometer sets a reference voltage v_r which is matched in the feedback loop with the residual motor armature v_a. The Momberg circuit has progressively flatter speed-torque curves as the control is reduced until the drive becomes unstable. Hence, the range is limited.

One of the many wide-range, feedback-type, half-wave circuits, as devised by Gutzwiller, is shown in Figure 6.4.[2] The circuit uses a zener diode reference and the motor residual voltage as a speed indication. The regulation is improved over the open-loop drives, but the circuit cost is higher than the three previously shown circuits.

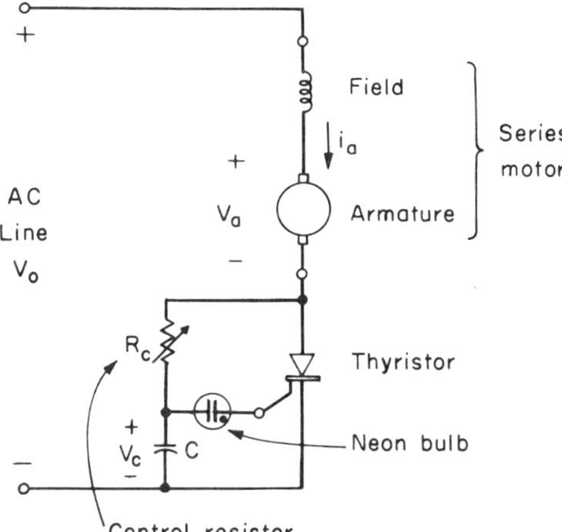

Figure 6.1   Half-wave series universal motor circuit.

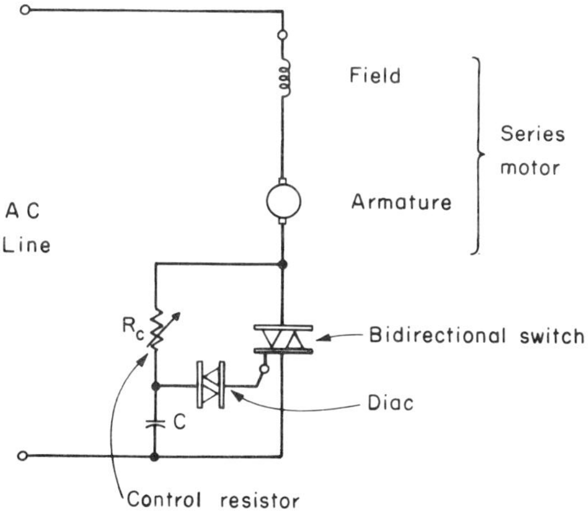

Figure 6.2   Full-wave series universal motor circuit.

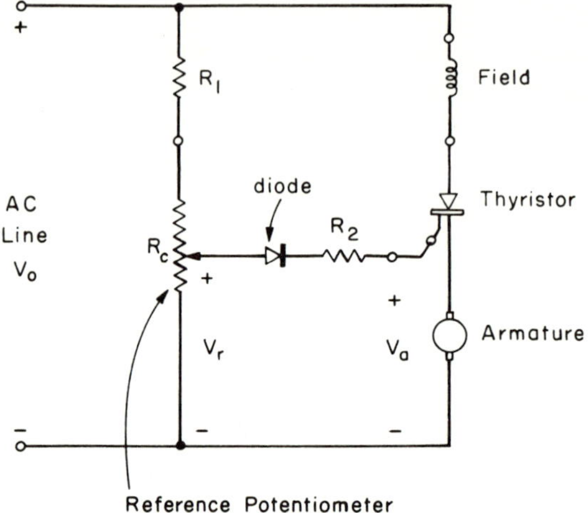

Figure 6.3   Momberg half-wave feedback circuit.

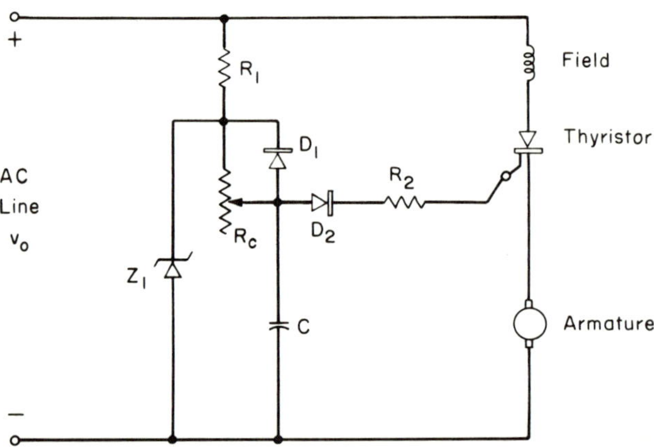

Figure 6.4   Gutzwiller half-wave feedback circuit.

6.2   *Operation*

The basic difference between the operation of the series motor circuits and the shunt motor circuits of Chapter 4 is that in the series motor, the field is produced by the armature current, whereas the field is independently supplied in the shunt motor. Hence, the armature generated voltage v_a is proportional to the speed and the armature current in the series motor, but only to the speed in the shunt motor.

The waveforms for the half-wave series motor circuit of Figure 6.1 are shown in Figure 6.5. They can be compared with the waveforms for the

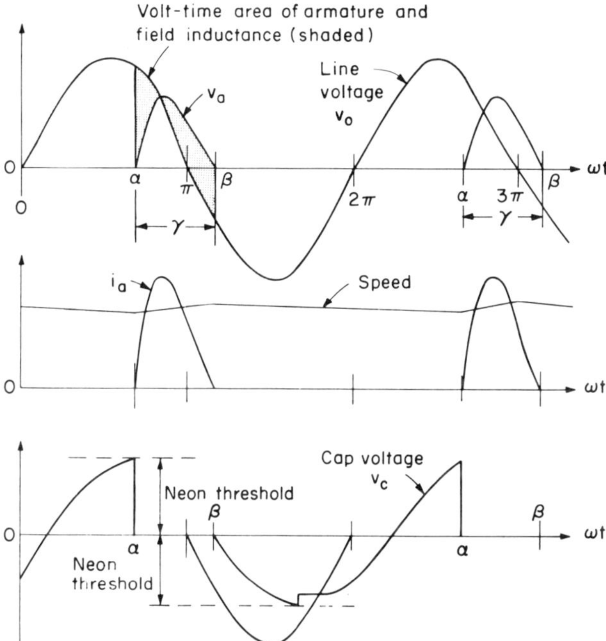

Figure 6.5   Waveforms for half-wave series motor drive.

shunt motor drive in Figure 4.3. Current flows in the field, armature, and thyristor of the circuit in Figure 6.1 when the thyristor is fired at $\omega t = \alpha$, and continues for a conduction angle γ until the thyristor blocks at $\omega t = \beta$. The current falls to zero only after the stored energy in the field and armature inductance is returned to the circuit, that is, the volt-time area balances to zero. During the conduction period α to β, the armature receives energy from the line and accelerates. From $\omega t = \beta$ to $\alpha + 2\pi$, the motor coasts and supplies the load torque from the decrease of its kinetic energy.

The armature generated voltage v_a is seen to be a replica of the current i_a. Actually, this is so because the speed is practically constant during the conduction interval. In an actual motor, the magnetic structure tends to saturate during the current pulse and the armature voltage becomes more flat topped.

The control circuit operation is shown in Figure 6.5. At $\omega t = \beta$, the capacitor starts to charge negatively through resistor R_c until its voltage reaches the threshold of the neon bulb. The bulb fires into the gate but the thyristor cannot conduct with negative anode-to-cathode voltage. The neon bulb holds the capacitor at its stable value until the line voltage starts to increase positively. The capacitor now starts to charge positively at a rate determined by resistor R_c. The voltage reaches the neon bulb threshold at $\omega t = \alpha$; the bulb discharges the capacitor into the gate terminal; the thyristor fires and places a low impedance from gate to cathode and from anode to cathode so that the capacitor voltage is clamped at zero until the conduction period ends. Increase of the control resistance R_c makes the thyristor fire later in the half-cycle resulting in reduced torque and motor speed. The earliest that the thyristor can be fired is when the line voltage v_0 equals the threshold voltage.

The waveforms for the full-wave circuit shown in Figure 6.2 are similar to those for the half-wave circuit. The full-wave circuit using the gated switch passes alternating current. A current pulse having negative polarity occurs between $\omega t = \pi$ and 2π. The negative current pulse produces negative magnetic field; since the torque is proportional to the product of current and field, the torque pulse is positive. The resultant torque pulses occur at twice line frequency in the positive direction. In the maximum conduction condition, the motor operates as though the ac line voltage were applied directly to the motor terminals.

The Momberg feedback circuit operates like the half-wave circuit of Figure 6.1 once the thyristor has fired. However, the firing signal for the thyristor is generated in a comparison circuit which gives the Momberg circuit speed regulating properties. The regulating portion of the circuit of Figure 6.3 has been redrawn in Figure 6.6b to show the operation.

During the times that no current flows in the armature and field winding, the residual magnetism of the field structure produces a small voltage across the armature terminals. The voltage is usually less than five volts and is directly proportional to the motor speed. The residual voltage is used in the Momberg circuit as a measure of motor speed. As shown in Figure 6.6b, the residual voltage v_a is compared with a reference voltage

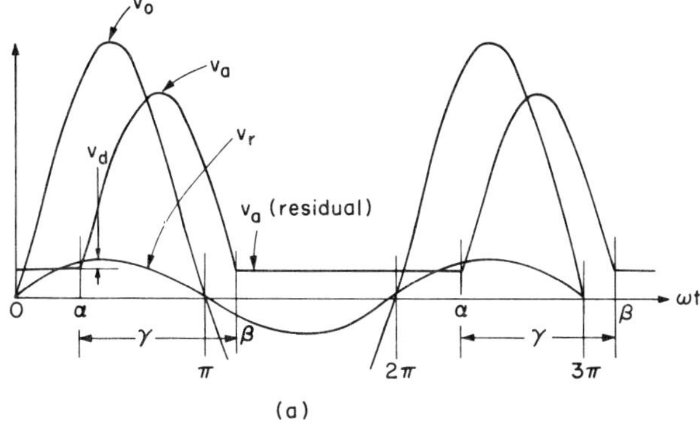

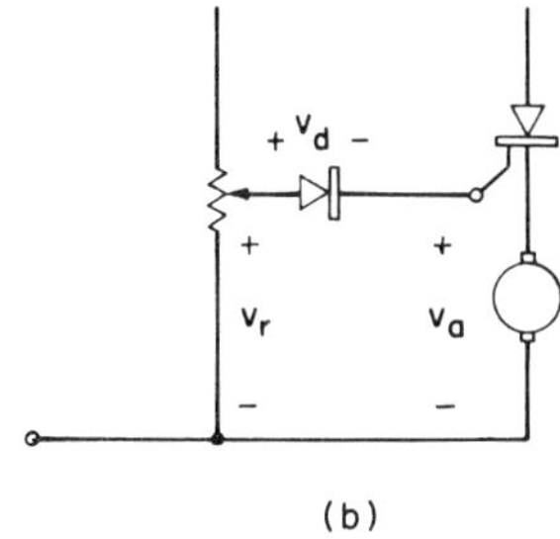

Figure 6.6   Momberg circuit: (a) waveforms; (b) circuit.

v_r; when the reference voltage less diode drop $v_r - v_d$ reaches the residual voltage v_a, the diode carries current and fires the thyristor.

The waveforms of the voltages are shown in Figure 6.6a. The thyristor is firing at α each cycle. Once the thyristor fires, the current increases the magnetic field and the armature voltage rises rapidly to the normal range of values. If the motor speed declines, the residual voltage declines and the thyristor is fired earlier in the cycle to restore the speed. The limitation of the circuit is that the comparison loop becomes more sensitive as the firing angle approaches $\alpha = \pi/2$. For reference voltage settings less than this value, the thyristor tends to fire once per several cycles and the motor speed tends to become erratic. Phase shift capacitors added to the reference branch help extend the speed range downward. A set of typical torque-speed curves is shown in Figure 6.7. Speed regulation is obtained over part of the range where the circuit has its maximum sensitivity.

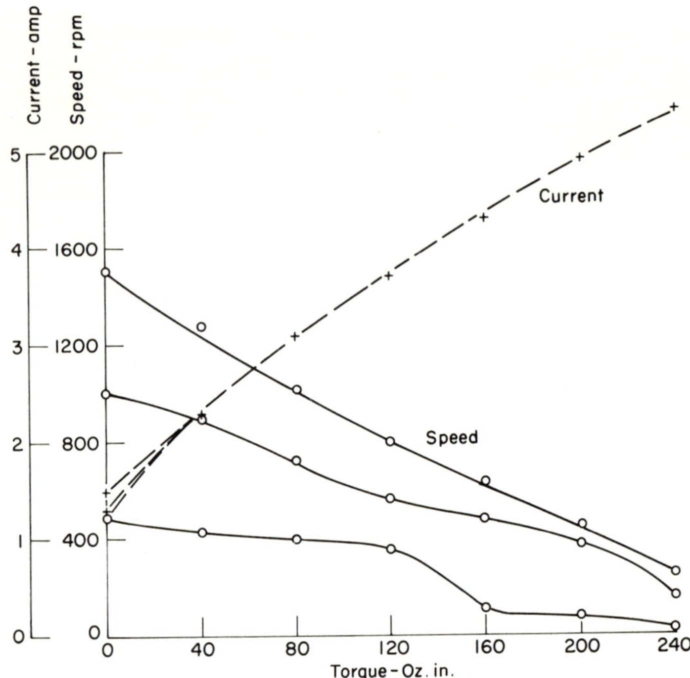

Figure 6.7   Measured speed and current versus torque
curves of Momberg circuit.

6.3 *Analysis, Phase Control – Unsaturated*

The speed-torque and current-torque characteristics of the series motor
under phase control are different from those with series impedance or
variable voltage control. Phase control of the semiconductor device not
only regulates the voltage applied to the motor, but determines the
conduction time each cycle during which current flows and the motor
field is excited. Since the internal loss torque frequently exceeds the useful
load torque, measured load torque-speed characteristics tend to disguise
the true behavior of the speed-control mechanism.

In this chapter, the characteristics of the series motor are derived as a
function of the firing angle of the semiconductor device to show how the
speed control is obtained. The operating region is divided into one zone in
which the motor current is significantly less than rated and the magnetic
structure of the motor is unsaturated, and a second zone in which the
magnetic structure is saturated. For portable tools such as electric drills,

the high internal torque losses of the fans, gears, bearings and brushes, and certain electrical losses, force the motor to operate almost wholly in the saturated region. The operation of the circuits will be treated for the half-wave case, but the final expressions will be given for both the half-wave and the full-wave cases.

The basic half-wave circuit of the thyristor and series motor is shown in Figure 6.8. The equation for the circuit when the thyristor is conducting is

$$v_m = L\frac{di_m}{dt} + v_a + Ri_m.$$ (6.1)

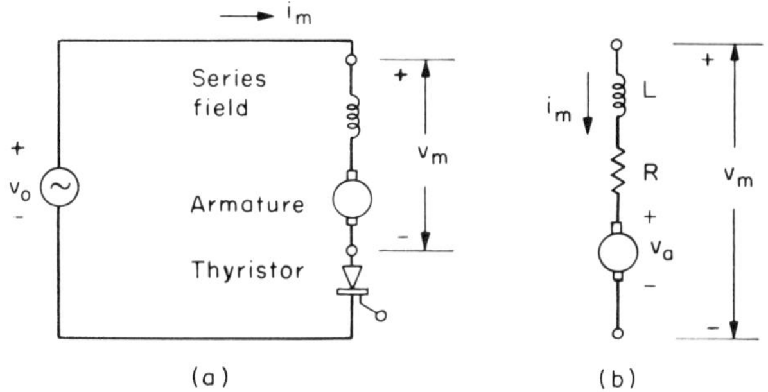

Figure 6.8   (a) Basic half-wave thyristor series motor circuit; (b) equivalent circuit of motor.

If the motor inductance L is not constant, then the equation should be written as

$$v_m = \eta\frac{d\phi}{dt} + v_a + Ri_m.$$ (6.2)

The armature voltage is given by

$$v_a = K_a N\phi.$$ (6.3)

The air gap torque for the series motor is

$$t = K_t \phi i_m.$$ (6.4)

For the motor in an unsaturated magnetic condition, the flux is proportional to motor current

$$\phi = K_m i_m.$$ (6.5)

The instantaneous torque, from Equations 6.4 and 6.5, is thus

$$t = K_t K_m i_m^2.$$ (6.6)

The average torque T is

$$T = \frac{K_t K_m}{\tau} \int_0^\tau i_m{}^2 \, dt. \tag{6.7}$$

However, the rms motor current I_m is defined as

$$I_m = \left[\frac{1}{\tau} \int_0^\tau i_m{}^2 \, dt \right]^{1/2}. \tag{6.8}$$

So that the average torque is given by

$$T = K_t K_m I_m{}^2. \tag{6.9}$$

Equation 6.9 is valid for the unsaturated motor for either half-wave or full-wave operation.

For low currents, the impedance voltage terms of Equation 6.1 are small compared to the armature voltage, and can be neglected

$$v_m \approx v_a. \tag{6.10}$$

Using Equations 6.3, 6.4, and 6.5, we find that

$$v_m = K_a N\phi = N K_a \sqrt{\frac{K_m}{K_t}} \sqrt{T} \tag{6.11}$$

as the instantaneous torque-speed-voltage relationship. If both sides of Equation 6.11 are squared and averaged, we obtain

$$\frac{1}{\tau} \int_0^\tau v_m{}^2 \, dt = \left[K_a \sqrt{\frac{K_m}{K_t}} \right]^2 N^2 T. \tag{6.12}$$

We assume that the speed remains constant over a cycle. Using the rms motor voltage V_m, Equation 6.12 becomes

$$V_m = \left[K_a \sqrt{\frac{K_m}{K_t}} \right] N \sqrt{T}. \tag{6.13}$$

Equation 6.13 gives the speed-torque relationship for the unsaturated series motor at low current for any waveform in terms of the rms motor voltage V_m.

The rms motor voltage can be expressed as a function of firing angle for half-wave control as

$$V_m = \left[\frac{\omega}{2\pi} \int_{\alpha/\omega}^{\pi/\omega} 2V_0{}^2 \sin^2 \omega t \, dt \right]^{1/2}$$

$$= V_0 \left[\frac{(\pi - \alpha)}{2\pi} + \frac{\sin 2\alpha}{4\pi} \right]^{1/2}. \tag{6.14}$$

6.4 *Analysis-Phase Control – Saturated*

In the high-current zone, we assume that the magnetic structure of the motor is saturated over the time of each current pulse. The instantaneous and average torques are given by

$$t = K_t \phi_s i_m,$$ (6.15)

$$T = K_t \phi_s I_m.$$ (6.16)

The armature voltage when the thyristor is conducting is

$$v_a = K_a N \phi_s.$$ (6.17)

To a first approximation we can neglect the inductance voltage of Equation 6.2 because the magnetic circuit is saturated, leaving

$$v_m = v_a + R i_m.$$ (6.18)

Equation 6.18 can be averaged over a half-period to yield

$$V_m = \frac{2}{\tau} \int_0^{2/\tau} v_a \, dt + R I_m.$$ (6.19)

Because of the saturation, the voltage v_a is of constant amplitude during the time of conduction. For a firing angle α, the half-wave average value is

$$V_a = \frac{K_a N \phi_s (\pi - \alpha)}{2\pi}.$$ (6.20)

Substituting Equation 6.20 into 6.19 yields

$$V_m = \frac{K_a N \phi_s (\pi - \alpha)}{2\pi} + R I_m.$$ (6.21)

Using Equation 6.16, to eliminate the current leads to

$$V_m = \frac{K_a N \phi_s (\pi - \alpha)}{2\pi} + \frac{RT}{K_t \phi_s}.$$ (6.22)

The motor voltage is also a function of α, because of the phase control of the thyristor

$$V_m = \frac{\omega}{2\pi} \int_{\alpha/\omega}^{\pi/\omega} \sqrt{2} V_0 \sin \omega t \, dt$$

$$= \frac{\sqrt{2} V_0}{2\pi} (1 + \cos \alpha).$$ (6.23)

Substituting Equation 6.23 into 6.22 and solving for the speed N leads to

$$N = \frac{\sqrt{2}\,V_0}{K_a\,\phi_s}\,\frac{(1 + \cos\alpha)}{(\pi - \alpha)} - \frac{2\pi R}{K_a\,K_t\,\phi_s^2}\,\frac{T}{(\pi - \alpha)}. \qquad (6.24)$$

It can be shown that the counterpart for full-wave phase control is

$$N = \frac{\sqrt{2}\,V_0}{K_a\,\phi_s}\,\frac{(1 + \cos\alpha)}{(\pi - \alpha)} - \frac{\pi R}{K_a\,K_t\,\phi_s^2}\,\frac{T}{(\pi - \alpha)}. \qquad (6.25)$$

6.5 *Phase Control – Interpretation*

Equations 6.24 and 6.25 show the speed-torque characteristics of the saturated series motor under phase control. The first term is the no-load speed and the second term is the speed droop as the motor is loaded with average torque T. Both terms are functions of the firing angle α; the motor behaves as though it were subjected to both voltage control and impedance control.

The behavior of the two terms of Equation 6.24 is shown in Table 6.1.

Table 6.1 Dependence of No-load Speed and Speed Droop on Firing Angle α. Saturated Series Motor.

Firing Angle α	$(1 + \cos\alpha)$	$(\pi - \alpha)$	No-load Speed $(1 + \cos\alpha)/(\pi - \alpha)$	Speed Droop $1/(\pi - \alpha)$
0	2.00	3.14	0.635	0.319
30°	1.87	2.62	0.713	0.382
60°	1.50	2.09	0.719	0.478
90°	1.00	1.57	0.638	0.637
120°	0.50	1.05	0.476	0.950
150°	0.13	0.52	0.257	1.92
180°	0	0	0	∞

The no-load speed, as angle α is retarded from zero-to-180°, first increases to 60°, then falls slowly to 90°, before declining significantly to 150° and beyond. The speed droop increases continuously as α is retarded.

Consider a typical operating range of $\alpha = 30°$ to $\alpha = 150°$. The no-load speed term varies by a factor of $0.713/0.257 = 2.77$; but the droop at a given value of torque varies by a factor of $1.92/0.382 = 5.03$. The droop term provides more speed variation with angle α than the no-load speed term.

The effect of the firing angle α on the droop term is equivalent to making the circuit resistance increase as $R/(\pi - \alpha)$. The average current I_m to produce the torque T is being squeezed into a smaller conduction angle as the firing angle is retarded toward 180°. The current must have a higher

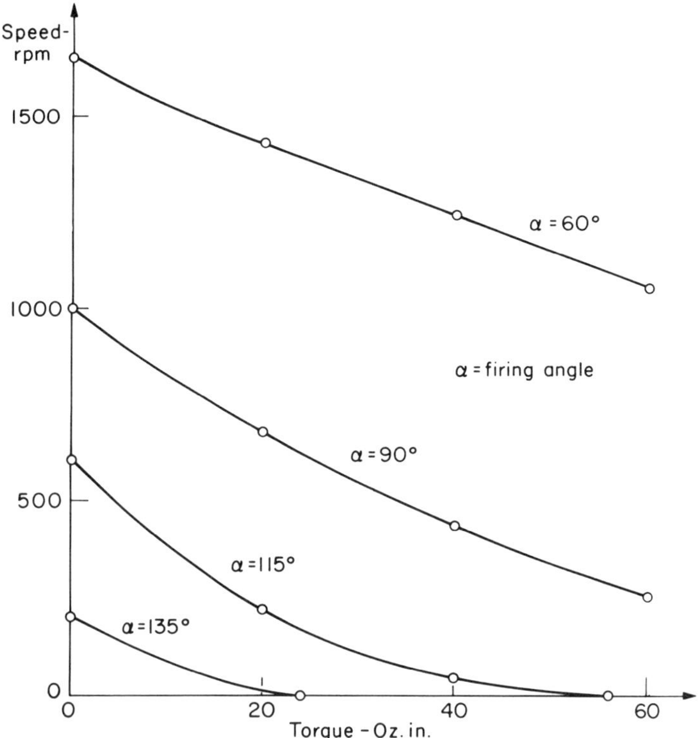

Figure 6.9   Measured speed-load torque characteristics of series motor and half-wave thyristor control.

peak to preserve its required average value, hence it produces a larger IR drop in the circuit resistance and a larger drop in the motor speed.

A set of test results for a half-wave controlled series motor in an electric drill is shown in Figure 6.9 for external load torque and in Figure 6.10 for internal air gap torque. The increasing steepness of the curves as the firing angle is increased is evident. It is obvious that the major part of the speed variation is coming from the second term of Equation 6.24, rather than the first term.

The analysis shows that phase control of the saturated series motor produces speed control through a combination of no-load speed variation and variation of the equivalent resistance of the circuit or speed droop as the firing angle is varied. In appliance and portable-tool motors, the torque losses force the motor to operate in the region in which the magnetic circuit is saturated and the analysis for the high-current zone applies.

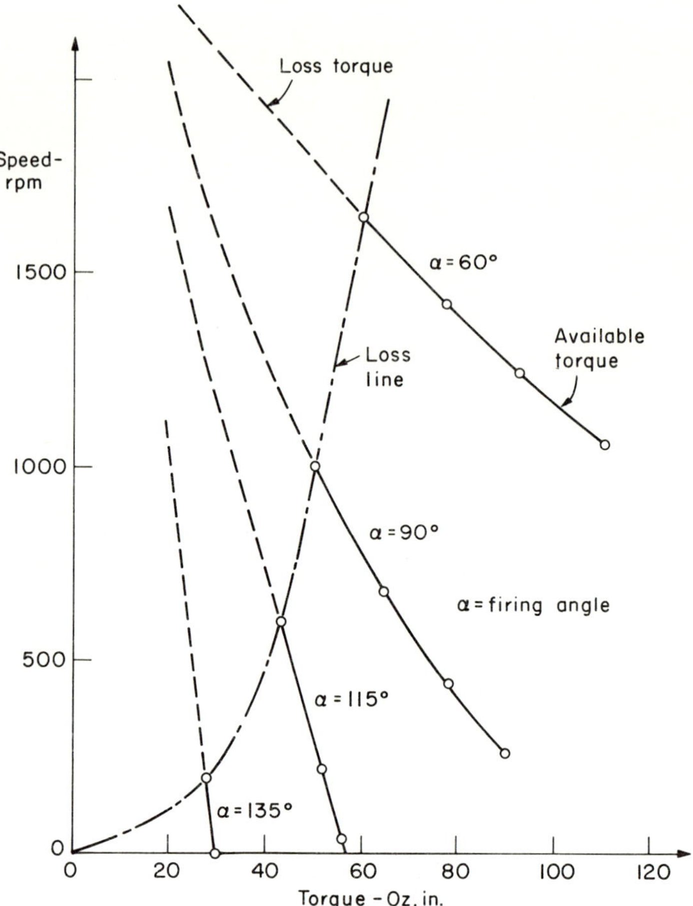

Figure 6.10   Speed-internal torque characteristics of series motor and
half-wave thyristor control.

6.6 *Summary*

Series universal motor drives usually consist of a simple thyristor circuit
operated as a half-wave controlled rectifier. Size and cost are at a
premium; feedback control of speed using the residual armature voltage as
a speed signal is used where the expense of the additional parts is
warranted. Analysis of the motor is best carried out with the assumption
of magnetic-circuit saturation.

A description of the construction and performance of series motors is given by Veinott.[3] Various types of control circuits for half-wave and full-wave operation are shown by Adem.[4]

References

1. J. W. Momberg, "Motor control systems," U. S. Patent No 2,939,064, Issued May 31, 1960.
2. F. W. Gutzwiller, "Universal motor speed controls," Auburn, N.Y.: General Electric Co., Application Note 200.4, June 1961.
3. C. F. Veinott, *Fractional Horsepower Electric Motors.* Second edition. New York: McGraw-Hill Book Company, Inc., 1948.
4. A. A. Adem, "Speed controls for universal motors," Auburn, N.Y.: General Electric Co., Application Note 200.47, June 1966.

7 DC-DC DRIVES

An important category of solid-state dc drives is that in which energy is obtained from a dc source such as a battery or overhead trolley wire. The circuits used are clever and unique for the application. They are referred to as *chopper circuits* when the solid-state device is switched at a high-frequency rate. Power transistors are used in the dissipative mode for lower power levels to drive torque motors.

7.1 *Application*

Chopper drive circuits are used for applications in which a dc series motor is operated from a dc source. The chopper replaces the series speed control resistors, which are wasteful of power. Variations of the chopper circuit can be used to provide regenerative braking of the motor or the return of energy back to the battery.

Typical applications of the chopper circuits are for electric battery vehicles, particularly fork-lift trucks, interurban and trolley cars, and marine hoists. The electric automobile will probably use a chopper circuit for speed control and braking. The obvious features of the chopper are smooth control, conservation of battery energy, and regeneration of battery energy.

7.2 *Principle*

The chopper speed control circuit is used with a series motor because the inductance of the field winding is required to maintain the armature current. A simplified circuit is shown in Figure 7.1a.

The principle of operation is that the series motor is supplied with pulses of source voltage whose average value is controlled by the ratio of on-to-off time of the pulses. For example, if the ratio is 0.5, the motor receives an average voltage of 0.5 V_0 and behaves as though it were connected to a continuous source of this value. Hence, at constant torque load, the motor speed will be proportional to the pulse ratio. Moreover, the control system does not dissipate power as a resistor would.

A thyristor is used to connect and disconnect the motor from the source, that is, to chop the source voltage. The thyristor requires a forced

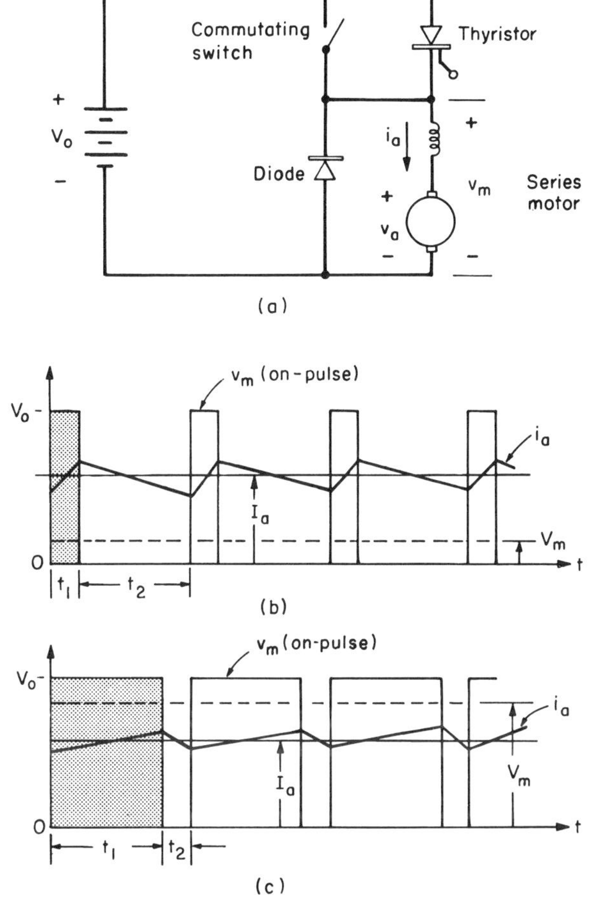

Figure 7.1   Chopper speed control: (a) circuit; (b) waveforms at low speed; (c) waveforms at high speed.

commutation circuit to reduce its current to zero and hold it long enough for the thyristor to revert to its blocking state at the end of each pulse. A capacitor and an auxiliary thyristor are used for this purpose.

The simplified circuit of Figure 7.1a shows the basic parts: the thyristor for switching the source to the motor; the freewheeling diode for carrying the armature current during the off periods; and the commutating switch for by-passing the thyristor current and turning it off. The thyristor is fired to start the pulse and the commutating switch closed to end the pulse. A separate circuit is required to generate the pulses and control the pulse ratio in accordance with the average motor current or speed.

The waveforms for two speeds at the same current and torque are shown in Figure 7.1b and c. For slow-speed operation, the pulse ratio and average motor voltage V_m are low. The average current I_a is determined by the torque requirement. The motor can receive energy from the source only during the on-time t_1. During this time, the voltage $(V_0 - v_a)$ is applied to the $R_a - L_a$ armature circuit and the current increases exponentially. During the off-time, the current decreases exponentially under the voltage $-v_a$ applied to the same $R_a - L_a$ circuit. The average current I_a for a pulse ratio k_r is given by

$$I_a = \frac{V_0 k_r - V_a}{R_a}. \tag{7.1}$$

At the same current I_a, the armature generated voltage V_a is proportional to speed. Hence, as k_r is adjusted, the motor speed follows suit at the same torque and current.

The waveforms for high-speed operation are shown in Figure 7.1c. Since the current i_a falls only a small amount during the off-time t_2, the voltage difference $(V_0 - v_a)$ need be small to increase the current during the on-time t_1. The speed of the motor thus rises to bring the voltage v_a close to V_0.

At constant pulse ratio k_r, the motor has the characteristics of voltage control as shown in Figure 2.6. It does *not* have the characteristics of resistance control because the average motor voltage is independent of the current. The circuit should have a current-regulating loop to limit the maximum current during acceleration. This requirement is particularly critical because the commutation circuit is designed to handle current up to a specific value. For excessive current, the thyristor will not turn off and control will be lost.

7.3 *Chopper Circuit*

A practical form of the chopper circuit for series motor control is known as the *Jones Circuit* and is shown in Figure 7.2.[1] The circuit uses a capacitor C, an autotransformer T_1 and a thyristor Tr_2 to perform the commutation function.

The operation of Figure 7.3 and the waveforms of Figure 7.4 show the somewhat complicated operation of the circuit over one cycle of on- and off-time. The basic problem of the design is to charge the capacitor C with enough energy and of the correct polarity so that when the thyristor

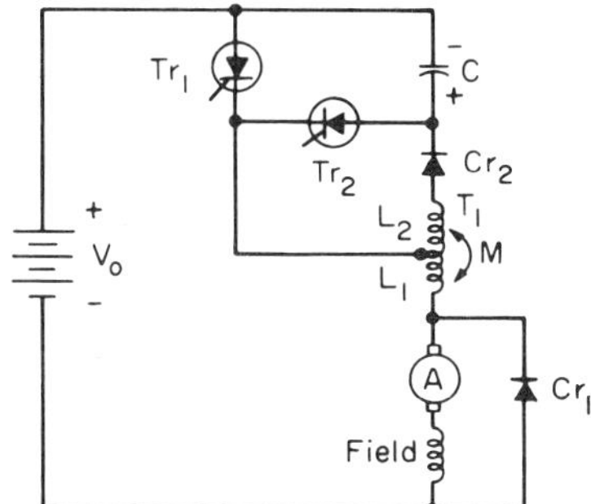

Figure 7.2   Basic Jones chopper circuit.

Tr_2 is fired, the capacitor will shut off Tr_1 and then Tr_2 as well. In the conducting or on-state, the capacitor is charged to about $-V_0$ volts and the motor current is carried by Tr_1 and the lower section of the autotransformer.

The detailed operation of the circuit can be followed from the sequence shown in Figure 7.3, where the thyristors are represented by switches. The corresponding intervals are shown in Figure 7.4.

Prior to time t_0, the motor current is circulating through the freewheeling rectifier Cr_1, the two thyristors are both off, and the capacitor is charged to an initial voltage with the lower plate negative. The on-interval is started by firing thyristor Tr_1. During the interval t_0 to t_1, the current I_0 to the motor produces through the autotransformer T_1 a current I_c which resonantly charges the capacitor to a peak positive voltage on the lower plate. Diode Cr_2 blocks and isolates the capacitor; motor current I continues as the source voltage V_0 is applied to the motor. This interval from t_1 to t_2 is the on-time of the motor.

To end the on-time, the commutating thyristor Tr_2 is fired at time t_2. First, the capacitor is clamped across Tr_1 applying a reverse blocking voltage to it and turning it off; second, the capacitor discharges through Tr_2 and resonantly recharges with opposite polarity; third, the Tr_2 current is forced to zero at t_4 so that it also blocks; fourth, the motor current transfers to the freewheeling diode Cr_1.

The capacitor voltage at t_4 exceeds the source voltage V_0, so the capacitor in the interval t_4 to t_6 discharges resonantly again back to the source and comes to rest when Cr_2 blocks at the level shown. This capacitor voltage is the level available at the start of the next on-interval. An additional thyristor can be used to charge the capacitor back to the source voltage between t_6 and the next t_0.

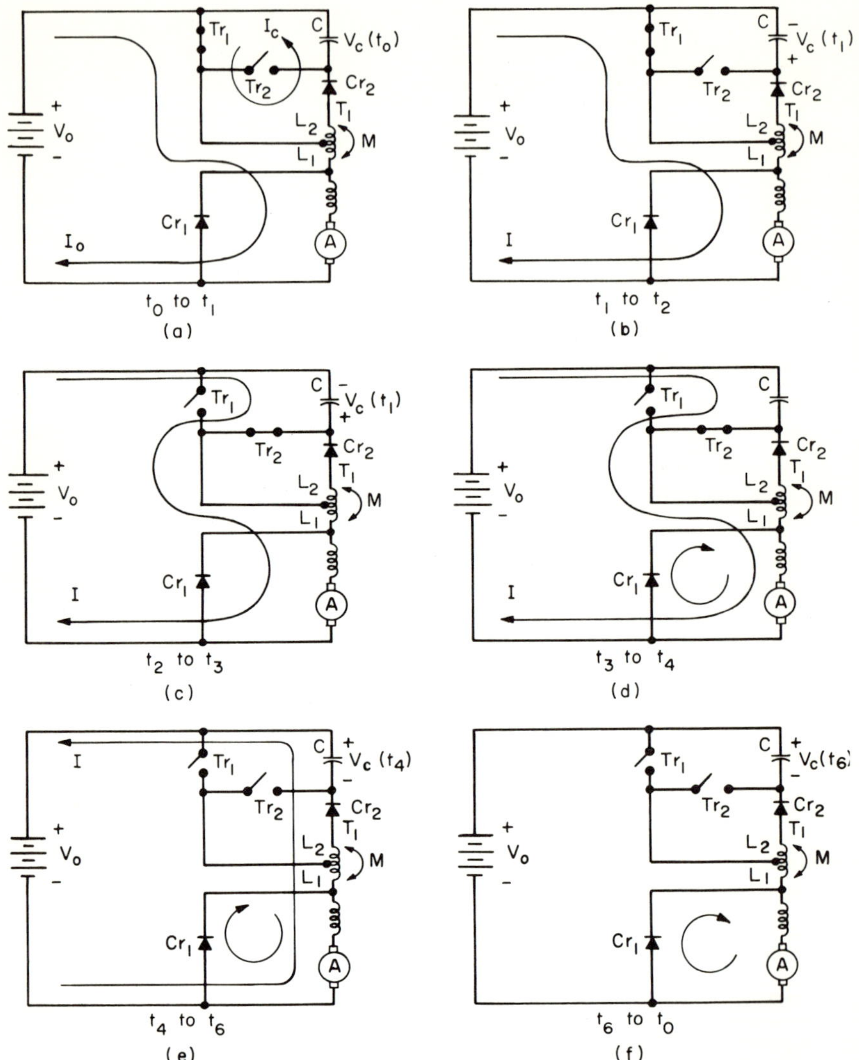

Figure 7.3   Modes of Jones chopper circuit during cycle of operation.

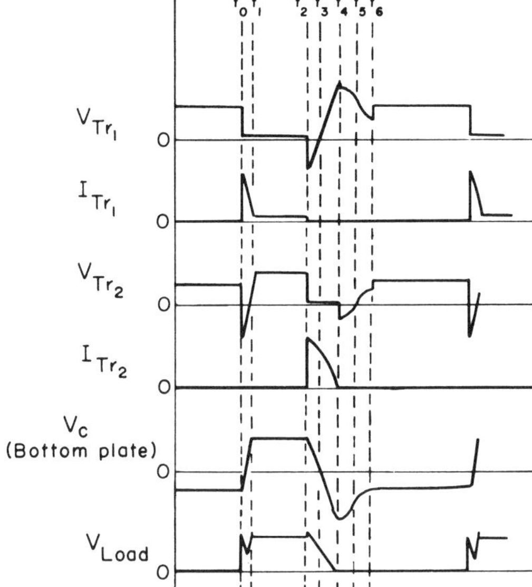

Figure 7.4   Waveforms during cycle.

Many other schemes have been developed for performing the pulse-ratio control function. In addition, various types of pulse modulation are used to vary the average motor voltage.

7.4 *Regeneration*

A feature of the chopper speed control circuit is that not only is it efficient for controlling the speed of the motor, but it can be used to brake the motor by returning the motor energy to the battery or any other dc source. A simplified circuit for regeneration is shown in Figure 7.5, which requires the addition of a diode and reconnection of the armature from the circuit of Figure 7.1.

The principle of regeneration operation is the use of the inductance of the field winding to store the energy of the motor when it is switched off the line, and the return of both the energy in the inductance and the direct output of the armature when it is switched on the line. The principle is analogous to that for speed control where the inductance is used to store energy when the motor is connected to the line and the energy is delivered to the armature when the motor is switched off the line.

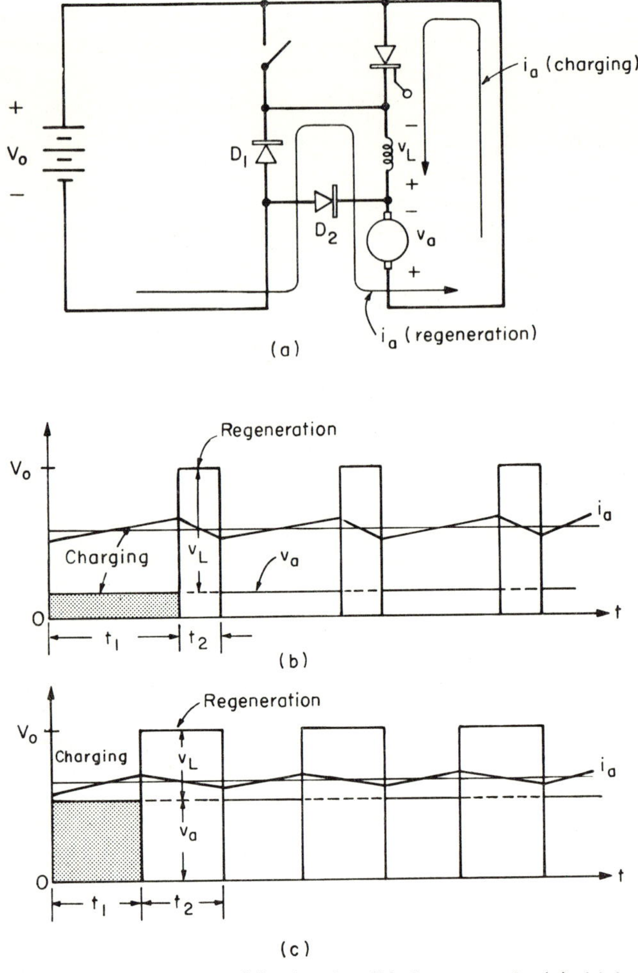

Figure 7.5   Regeneration: (a) circuit; (b) low speed; (c) high speed.

During the charging interval in Figure 7.5, the thyristor is conducting and the armature voltage v_a is applied to the armature $R_a - L_a$ circuit causing the current i_a to rise exponentially as the inductance absorbs energy. The interval is shown as t_1 in Figures 7.5 b and c. Meanwhile diodes D_1 and D_2 are blocked, if the armature voltage $v_a < V_0$. When the current i_a has risen to a prescribed value, the thyristor is commutated off and the armature current transfers to the regeneration path through diode D_1.

During the regeneration interval t_2, energy is transferred back to the battery. The voltage difference $(V_0 - v_a)$ is now applied to the armature $R_a - L_a$ circuit in a direction to decrease the current i_a exponentially. The current falls until the end of the interval when the thyristor is fired to start charging the field inductance again. The torque direction in the motor is the braking direction which will extract mechanical energy from the load and the connected inertia.

The waveforms for two values of speed at the same braking torque and average current are shown in Figure 7.5 b and c. For slow-speed operation, the ratio of thyristor conducting time to total time, which is the pulse ratio k_r, is high. The long time t_1 is required to charge the inductance from the low armature voltage and the short time t_2 to deliver the energy to the line.

For high-speed operation, a shorter charging period t_1 and a longer regeneration period t_2 are required to transfer the increased energy to the dc source. The pulse ratio is controlled by monitoring the average armature current I_a during the braking period.

7.5 *Summary*

The chopper circuits are used to obtain efficient speed control and regeneration of a series motor operated from a battery or other dc source. A thyristor, which must be force commutated, is used to connect and disconnect the motor from the source. During the disconnect intervals, the field winding inductance is used as an energy reservoir.

An original paper on chopper speed-control circuits was presented by Jones.[2] Examples of applications are given by Heumann,[3] and more recently by Beasley and White,[4] and Zeccola and Weiser.[5]

References

1. N. W. Mapham and J. C. Hey, "The control of battery powered dc motors using SCR's in the Jones circuit," *IEEE International Convention Record*, pt. 4, pp. 124–134, 1964.
2. D. Jones, "Variable pulse width inverter," *Electron. Equipment Eng.* pp. 29–30, November 1961.
3. K. Heumann, "Pulse control of dc and ac motors by silicon controlled rectifiers," *1963 Proceedings of the Intermag Conference*, Washington, D. C.
4. J. Beasley and G. White, "The use of thyristors for the control of a dc traction motor operating from a 600 volt line supply," *Proceedings of the 1965 IEE Conference on Power Applications of Controllable Semiconductor Devices*, pp. 187–195.

5. R. A. Zeccola and E. F. Weiser, "Thyristor (SCR) chopper control system for transportation equipment," *IEEE Conference Record of IGA Third Annual Meeting,* pp. 471–482, 1968.

8 SPECIFICATION OF DRIVES

The description of a drive system requires the specification of a number of physical and performance features. This description requires a terminology which can be understood alike by designers, manufacturers, and customers. This chapter represents a dictionary of such terminology.*

8.1 *Speed*

1. *Speed Regulation.* Regulation is the ability of the drive to maintain the preset speed under varying loads. At any preset speed, as additional load is added, the speed of a dc shunt motor tends to fall off. The regulator in the drive circuit detects the speed drop and compensates for it by increasing the armature voltage, thus bringing the speed back to its preset level. Regulation is generally defined as the per cent change in speed from no load to full load (or to close to full load, such as 95 per cent of full load). Essentially the regulation percentage describes the slope of a drive's speed-torque characteristic and, therefore, is a function of the characteristics of both the drive's control unit and the motor. Most manufacturers state regulation as a percentage of the base speed. For example, if the base speed is 3500 rpm and the regulation is 5 per cent of base speed, this means that at full load the speed is 3325 rpm (3500 − 175). But if the regulation is given as a per cent of base speed, at some speed other than base, the drive still drops in speed by the same number of rpm. Hence, if the 5 per cent drive is running at only 2000 rpm at no load, at full load it will drop in speed to 2000 − 175 or 1825 rpm. This is actually an 8.8 per cent change in set speed. Another thing to watch about regulation figures is that, if they are given simply at base speed, the per cent stated may not apply throughout the entire speed range. Thus a 5 per cent regulation at base speed may be a 7 per cent regulation at 1/3 of base speed.

* From A. Kusko, "Packaged SCR dc-drives," *Control Eng.* pp. 60–61, July 1966.

2. *Speed Range.* The speed range of a drive, coupled with the drive's base speed, defines the lowest speed at which the drive can be operated. Two factors must be considered in relation to low-speed operation. The first is the cooling of the motor, the second is the regulation of the drive. At low speeds, motor cooling is at a minimum; therefore, the torque conditions which apply must be known before a speed range figure has any true meaning. For example, a 20:1 range, stated at continuous rated torque, gives a clear indication of the true operating conditions of the drive. Simply to state that the speed range is 20:1 has no meaning. Under light load, the speed range is generally wider than under full load. But when the speed range is stated for the lightly loaded condition, the regulation of the drive is most important since any increase in load on the motor at low speed will tend to stall the motor. The essential difference between a speed range stated for continuous rated torque and one for the lightly loaded condition is that, for the former, speed droop of the drive is included in the range specification, whereas in the latter case droop is not included.

3. *Adjustable Maximum Speed.* The maximum speed at which the motor will run can be set by an adjustment screw or knob in the drive cabinet. Thus, although the speed setting potentiometer reads full speed, the motor will run slower.

4. *Adjustable Minimum Speed.* The minimum speed at which the motor will operate can be set by an adjustment screw or knob in the drive cabinet. Thus, although the speed setting potentiometer reads zero, the motor will run at some speed above zero.

5. *Base Speed.* Base speeds refer to the maximum speed at which the motor can run with full field applied. Most manufacturers offer several standard speeds while many also offer gear reducers for a wide choice of top speed.

6. *External Speed Reference.* The speed reference signal, normally derived from a potentiometer, can be supplied from an external voltage source. For example, if a slave drive is to follow the speed variations of a master drive, a tachometer, driven by the master, can supply the slave's reference.

7. *Preset Speeds.* By providing multiple speed-setting potentiometers and relays to switch in whichever potentiometer is required, speeds can be preset and selected by operating pushbuttons or by some other means of operating the relays.

8. *Speed Sensing, CEMF.* The counter emf voltage generated by a dc motor's armature is proportional to the speed at which the

armature is rotating; by detecting this voltage, a signal proportional to speed is obtained. This voltage is compared with the reference speed, any error between the two causing the drive circuits to increase or decrease the speed accordingly (*see IR*-Drop Compensation).

9. *Remote Speed Control.* If the operator's station is located remotely from the drive, the speed reference potentiometer is usually mounted in the station and the potentiometer signal extended to the drive through a three-wire circuit. Sometimes the reference voltage supply is also mounted in the operator's station and the speed reference voltage, selected by the potentiometer, is fed to the drive.

10. *Field Weakening.* The base speed of a drive can be extended upward by reducing the field current.

11. *IR-Drop Compensation.* When CEMF speed sensing is used, the armature terminal voltage is detected as a measure of the motor's actual speed. But the terminal voltage comprises two elements, the counter emf voltage generated by the armature, and the voltage drop in the resistance of the armature windings (the *IR*-drop). Since only the CEMF voltage is actually proportional to motor speed (the *IR*-drop voltage is proportional to armature current), by eliminating the *IR*-drop component from the terminal voltage, a more accurate indication of actual motor speed is possible. Thus *IR*-drop compensation is a means of subtracting a signal, proportional to the *IR*-drop, from the terminal voltage. But, because the *IR*-drop varies with different motors, the compensation must be set for a specific motor.

12. *Jog.* Jogging controls are used to operate the drive at very low speeds during setting-up operations. The jog control can be designed to run the drive while the pushbutton is held down, or it can be designed to run the drive continuously at very low speeds for threading operations.

8.2 *Reversing – Braking*

1. *Reversibility.* Most drives are unidirectional with reversibility as an added cost extra. Reversibility implies that the drive control has a switch or contactor to reverse the dc leads to the motor armature. Some drives also offer dynamic braking as a standard with the reversibility option.

2. *Regenerative Braking.* Regenerative braking, like dynamic braking, is a means for extracting mechanical motor energy electrically from the armature terminals. The electrical energy is pumped back into the supply line instead of being dissipated in a resistor. This form of braking allows the motor to be braked to standstill.

3. *Dynamic Braking.* Dynamic braking is a means of electrically loading the motor as a generator to extract its stored mechanical energy and reduce its speed. This is usually done by connecting a resistor across the armature terminals while maintaining full field excitation. The resistor is connected by an auxiliary contactor, or by a set of midposition contacts of a manual reversing switch. Dynamic braking is most effective when the motor is running at a high speed.

4. *Quick Slowdown.* By including dynamic braking in the drive, the motor slows down at a faster rate than it would if simply allowed to coast to a stop. (*see* Dynamic Braking).

8.3 *Acceleration*

1. *Controlled Acceleration by Current Limit.* By setting the maximum allowable armature current, the acceleration of the load can be kept down to a safe level on startup of the drive. Controlled acceleration of this type is inherent in all drives that have a built-in current limit.

2. *Controlled Acceleration by Timing.* The acceleration of the motor and load is controlled by delaying the full application of the speed reference signal. Thus, if the speed setting is changed abruptly by turning the dial of the speed setting potentiometer rapidly, the rate at which the speed changes is an automatically controlled function independent of the dial. With this feature, acceleration is controlled throughout the speed range and not only during startup, as in the case with current limit control. Electronic delay circuits or auxilliary motor-driven potentiometers are commonly used to delay the reference signal.

8.4 *Torque – Current Limit*

1. *Torque Limit.* This is essentially current limit where the torque output from the motor is kept down by limiting the armature current (*see* Torque Regulation).

2. *Torque Regulation.* Torque regulation (or torque control) means one of two things: simply an adjustable torque limit (current limit) where the drive maintains constant set speed out to a preset torque

and then maintains constant torque at a declining speed for further mechanical loads; or the incorporation in the drive of a particular drooping torque-speed characteristic which can be varied.

3. *Current Limit.* Armature current must be limited by the drive controller to protect both the motor and the SCR's from overheating due to excessive currents, and also to limit the torque applied by the motor to the load. The armature current is sensed either by detecting the voltage developed across a low resistance inserted in the armature leads, or by some form of dc transductor which senses the armature current. When the armature current reaches the set limit, speed regulation becomes ineffective and the motor speed will drop sharply with increased load until the load demand that caused the current to reach its limit is removed.

4. *Current Limit, Adjustable.* This is the same as current limit but the limiting current value can be adjusted to match the drive to a particular load and motor.

8.5 *Equipment*

1. *Disconnect Switch.* Drive packages incorporate a contactor or switch to start and stop the motor. An external ac disconnect switch or contactor is desirable to shut down the entire drive for safety, to allow for remote tripping, or to provide a way of fusing the line to the drive.

2. *Line Contactor.* When a drive package does not incorporate a contactor, but uses a manual start-stop switch, an ac line contactor can be added for local and remote pushbutton operation, for overload protection, and for interlocking.

3. *Supply Transformer.* This is a transformer which is mounted external to the basic drive package to provide the ac input voltage specified for the drive from a supply line of a different voltage. Most drives do not incorporate their own transformer because the dc motor voltage is matched to the rectified ac voltage.

4. *Thermal Relays.* Magnetic and thermal overload relays can be placed in either the ac or dc power circuits of the drive to protect either the SCR's or the motor. Although fast-acting fuses are frequently used for overload protection, better long-time protection can be obtained by using both relays and fast-acting fuses. The overload relays can also be adjusted for different operating conditions.

5. *Thyrector.* Trade name of the General Electric Co. for a back-to-back diode pair. Used in drives to protect the SCR's from high voltage transients.

6. *Triggering Circuit.* The triggering or firing circuit determines at which point in the positive half-cycle of the ac line voltage the SCR's will start to conduct (*see* SCR and Phase Control Rectification).

7. *50 HZ Option.* Magnetic components such as transformers, ac motors which are used for fan-cooling the drive cabinet, relays and magnetic amplifiers all require additional core material so that they will not saturate during 50 Hz operation. The speed of the ac motor is also reduced, thus decreasing the flow of cooling air. Therefore, a drive package designed for 60 Hz operation may not be satisfactory for 50 Hz supplies.

8. *Phase Control Rectification.* Basic principle on which all SCR dc drives operate. The point at which the SCR fires during the positive half-cycle of the ac line voltage applied across the SCR is determined by the angle from the start of the half-cycle at which the triggering signal is applied to the triggering terminal (*see* SCR). The SCR stops conducting when the current passes through zero. Thus the ac voltage is rectified to a dc voltage whose average value can be adjusted simply by changing the triggering angle. Advancing this angle toward the start of the positive half-cycle increases the average voltage, retarding the angle toward the end of the half-cycle decreases the average voltage. Changing the average voltage which is applied to the dc motor's armature varies the motor speed.

9. *Thyristor (SCR).* Silicon controlled rectifier, sometimes called a thyristor. An SCR is a three-terminal solid state device which operates like a thyratron gas tube. When a positive voltage is applied across the anode and cathode terminals and a positive signal applied to the trigger terminal (equivalent to the grid in tube terminology) the SCR "fires" and conducts current. Once fired, the trigger signal can be removed without stopping the conduction. Conduction stops when the positive voltage is removed from the anode-cathode terminals.

9 CONTROL EQUIPMENT

Control systems for solid-state dc drives include three essential portions: 1, thyristor firing circuits; 2, speed and current sensing means, and 3, reference, comparison, and amplifier circuits. There is a wide variation in the circuits and hardware used to accomplish these functions depending upon the size of the drive and the manufacturer.

9.1 *Firing Circuits*

Thyristor firing circuits are classified as magnetic firing circuits and solid-state firing circuits. Magnetic firing circuits use either magnetic amplifiers or adjustable saturable reactors to control the firing angle in accordance with a dc signal. Solid-state firing circuits use capacitors as the source of the firing energy and release it through some type of solid-state device which breaks down and conducts at a threshold voltage. The choice of one type over the other depends on the required range of control, the length of the pulse, and the relative cost.

An elementary magnetic firing circuit is shown in Figure 9.1a for a single-phase full-wave thyristor bridge. The magnetic amplifier is required to develop each half-cycle sharply rising voltages v_1 and v_2 across the amplifier resistors and on the gate terminals of the thyristors. The angles α_1 and α_2 of the voltages are required to be controllable by the control current I_c over the range, as nearly as possible, 0 to 180 degrees. Hence, as the control current is adjusted, the firing angle is adjusted and the motor speed is controlled.

The magnetic amplifier consists of two square-loop cores, designated I and II, upon which a gate winding and at least one control winding are placed. The two cores are supplied from independent secondary windings of an isolation transformer, because the thyristors are not connected with a common cathode. Each secondary winding of the transformer is typically rated 15V, one ampere.

The magnetic amplifiers operate on a volt-time area principle. In Figure 9.1b, during the first half-cycle, the control current I_c applies a negative volt-time area to core II, while its diode decouples its gate winding from the source. The core flux is driven down from saturation by an amount $\Delta\Phi$ by the time $\omega t = \pi$. In the second half-cycle, the diode of core II conducts

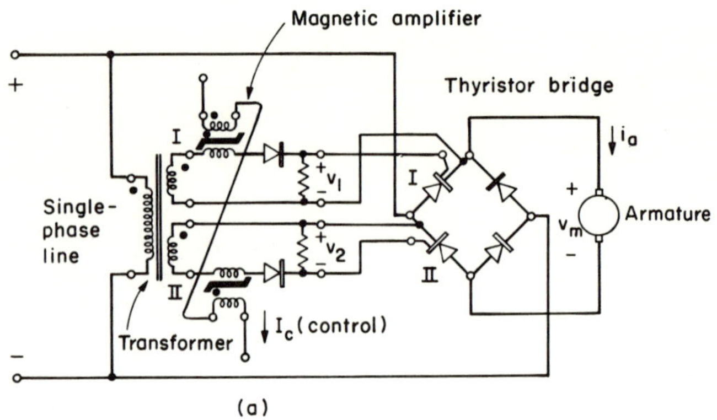

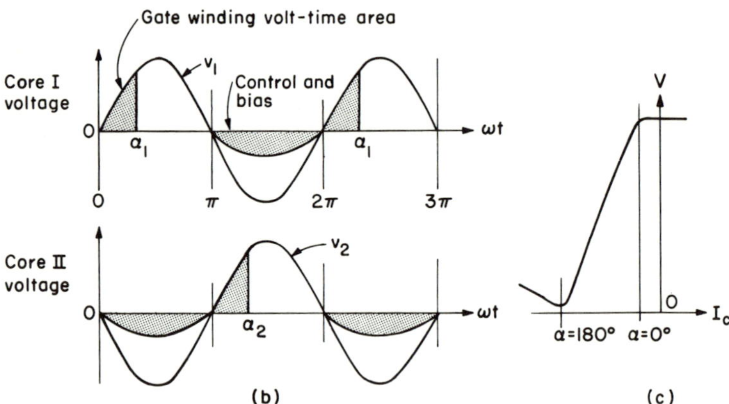

Figure 9.1 Single-phase magnetic firing circuit.

and the transformer voltage is applied to the gate winding driving the core flux back toward saturation. When the winding has absorbed the same volt-time area as the core received during the previous half-cycle, the core II saturates and applies the transformer voltage during the interval $\omega t = \alpha_2$ to 2π to the load resistor and the gate as voltage v_2. The sharply rising front on v_2 triggers thyristor II and causes line voltage to be applied to the motor over the same interval. The same phenomena take place on core I except that the saturation of the core and triggering of thyristor I occur on the alternate half-cycles. Hence, the motor receives a full-wave rectified and controlled voltage from the thyristor bridge.

The control characteristic is shown in Figure 9.1c. Control takes place for negative control current. The net control ampere-turns are usually provided by a bias winding and a control winding connected so that speed increase occurs for control current increase.

The magnetic amplifier firing circuit has two features. First, the firing circuit can be made to be responsive to several signals by introducing them into separate control windings. Second, the pulse delivered to the gate has a width of $(180° - \alpha)$ so that the thyristor will refire during the interval if required to do so. However, the magnetic amplifier does introduce a time delay that complicates the compensation design of a feedback system.

Solid-state firing circuits are built in varying degrees of complexity depending upon the means for control, the pulse shape, and the pulse power. The simplest type utilize a capacitor for energy storage, a charging circuit, and a voltage-sensitive trigger that discharges the capacitor into the gate circuit of the thyristor. The types that require a finite pulse width utilize a blocking oscillator or equivalent source that delivers a pulse in response to an initiating signal. A third type utilizes a high-frequency oscillator, say 10 kHz, and applies a burst pulse of 10 kHz voltage to the gate each time the thyristor must be fired.

A simple version of the solid-state trigger circuit is shown in Figure 9.2a. The circuit uses a unijunction transistor as the voltage-sensitive trigger. The unijunction breaks down when the emitter voltage E to B_1 reaches a fraction of the voltage between the bases B_1 and B_2. When breakdown occurs, the E to B_1 junction drops to a low resistance and allows the capacitor C to discharge through the pulse transformer.

The zener diode Z functions to clip the rectified voltage at a standard level and form the voltage v_1, which is applied to the charging circuit $R_c C$. The dipping of the voltage v_1 to zero at the start of each half-cycle serves to synchronize the trigger circuit with the voltage applied to the thyristors. The capacitor C charges by current i_1 at a rate set by R_c. The capacitor voltage is shown in Figure 9.2b as v_c. When the voltage v_c reaches the unijunction threshold, the $E - B_1$ junction breaks down and the capacitor discharges via current i_2 through the pulse transformer.

The secondary windings of the pulse transformer see the pulse voltage v_{p1} of Figure 9.2b. The two secondary windings would feed the same pulse to two thyristors of a full-wave circuit. Only the one with the positive anode-cathode voltage would fire. As soon as the capacitor discharges, it starts to recharge as shown.

Control is obtained by varying the charging resistor R_c to affect the slope of the capacitor voltage. The range of control is less than $180°$ and is of the order of $150°$. The pulses are narrow and have an energy limited by the discharge current rating of the unijunction. Where the circuit must be used in a feedback system, the firing angle α must be controlled by a dc

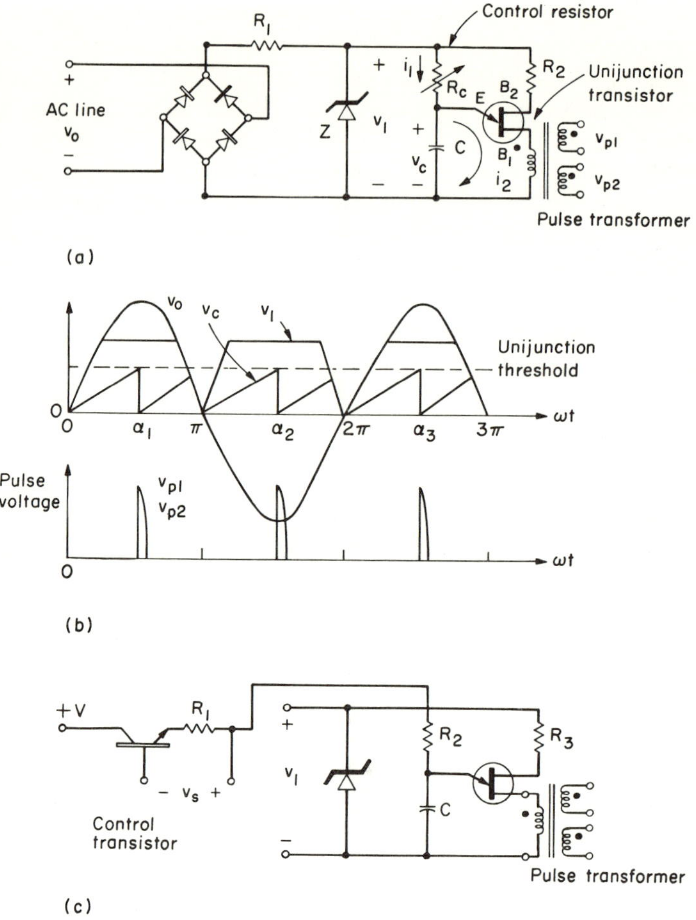

Figure 9.2 (a) Unijunction capacitor discharge firing circuit; (b) firing circuit waveforms; (c) firing circuit for operation from a control signal.

signal. The means for incorporating a drive transistor is shown in Figure 9.2c. The charging current is supplied from a separate source but the same form of synchronizing signals is maintained.

Three-phase versions of these circuits are obtained by supplying three individual circuits from suitable phases of the supply line. Because the $\alpha = 0$ point comes at $\omega t = 60°$ for complete bridges and at $\omega t = 30°$ for half-wave circuits, it is impractical to operate the firing circuit from the same voltage as the thyristor being controlled and waste the first 30° or 60° of range. It is practical to use a voltage that is advanced by 30° or 60° and obtain a full useful control range; hence, firing circuits are usually connected line-to-neutral for bridge circuits and line-to-line for half-wave circuits.

A three-phase magnetic firing circuit that operates from a single control transistor is shown in Figure 9.3.[1] The circuit uses the same principle as

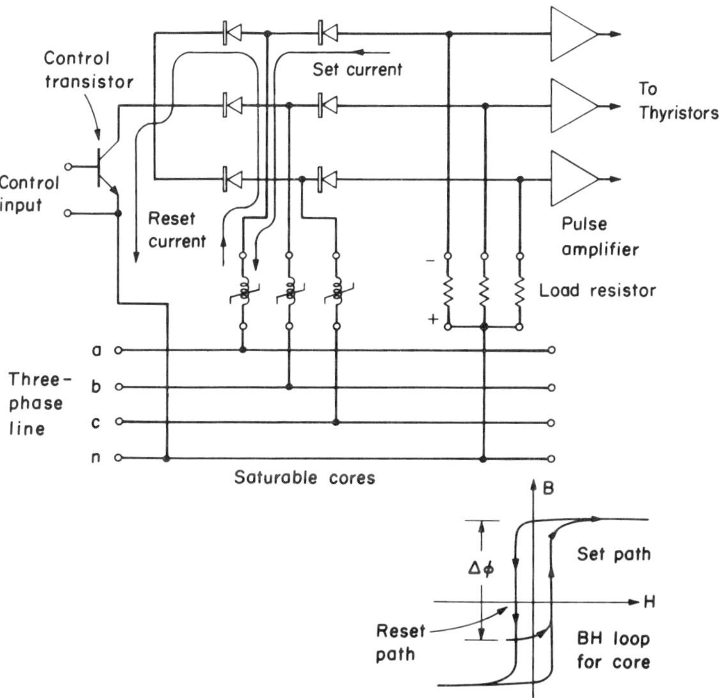

Figure 9.3   Three-phase magnetic firing circuit.

that shown in Figure 9.1 except that the core uses a single winding to perform both the control and gate function. The circuit is simple and provides about 160° of firing angle range.

The circuit operates from the three-phase line-to-neutral voltages. The operation can be seen by observing the a-phase core. When the voltage v_{an} is positive, a reset current passes through the winding, diode, and control transistor, resetting the core by $\Delta\Phi$ as shown. On the negative half-cycle of v_{an}, a set current flows; when the core has absorbed a volt-time area equal to $N\Delta\Phi$, the core saturates and allows the remainder of the half-cycle of voltage to appear on the load resistor. This voltage is amplified and applied with proper polarity to the thyristor gates.

As the control transistor is driven harder, the reset current increases and the cores are caused to saturate later in the negative half-cycle. That is, increased signal retards the firing angle.

The circuit shown in Figure 9.3 is suitable for providing three gate signals for the thyristors in an incomplete bridge. To develop six gate signals, six cores must be used and the circuitry becomes more complicated.

9.2 *Speed Sensing*

Except for some fractional-horsepower drives, all solid-state dc drives employ some feedback (closed-loop) means of speed control. The speed setting dial controls a reference voltage; this reference voltage is compared by the control system with a voltage proportional to speed; the difference is the error voltage which is amplified to control the firing circuits in a direction to minimize the error voltage. The speed regulating ability of the drive depends upon the fidelity of the speed sensing signal.

Two methods are used for generating the speed signal. The most common is to use the armature voltage under the condition of constant field current and to refine it by subtracting the IR drop, that is

$$V_a = V_m - I_a R_a. \tag{9.1}$$

The armature voltage is sampled with a resistance divider. The armature current is measured in one of the ways to be discussed and the subtraction carried out arithmetically in the feedback circuit, or in an operational amplifier. The multiplier representing R_a is set by test to include all resistance-like effects.

The regulating accuracy of armature voltage speed sensing is of the order of ± 2 per cent of the base speed. The signal is directly affected by the deviations of the field flux under saturation and line-voltage change, and by temperature, linearity, and setting of the armature-voltage circuit. The method is practical and inexpensive. If part of the speed range is covered by field weakening, then the method is not suitable if speed must continue to be regulated.

More accurate speed regulation is achieved by using a tachometer driven from the motor shaft. The tachometer is an ac or dc generator which has a high order of linearity between its speed and output voltage. For dc drive systems, dc tachometers are usually used. The dc tachometer is built with a permanent magnet field and sometimes with silver brushes to reduce the brush-to-commutator voltage drop. It is normally wound with a relatively high internal impedance so that it may require external temperature compensating resistors and a high impedance load. Typical voltages are 10 V per 1000 rpm.

Tachometers are available to secure regulating accuracies of ± 0.1 per cent of the base speed. To achieve such regulation the speed regulating control loop must have very high gain and may be difficult to stabilize over the operating range of the motor. The loading on the motor affects the parameters· of the control loop and affects the dynamics. The commutator of the tachometer produces a ripple on the output voltage. At high speeds, the ripple frequency can be easily filtered; however, at low speeds, the ripple may require a filter whose break frequencies fall into the control loop range and affect the dynamics of the system. Special large diameter tachometers, with a large number of commutator segments, are built to operate at low speeds.

The most accurate drives are built using pulse or digital speed signal and reference techniques. However, the designer should evaluate the use of digital speed control of a dc drive with the alternative of a frequency controlled synchronous ac drive, which can achieve high orders of steady-state speed regulation without requiring high feedback loop gains.

9.3   *Current Sensing*

A signal proportional to line or armature current is required on almost every dc drive system for the following reasons: 1, to develop the *IR* compensation signal when armature-voltage speed sensing is used; 2, to monitor possible misfiring of the thyristors in reversing drives; 3, to use for regulating current, torque, or acceleration when the motor is called upon to make changes in speed.

Three techniques are used for current sensing. They are 1, a resistance shunt in the armature circuit; 2, a dc transductor in the armature circuit; 3, current transformers in the ac lines to the controlled rectifiers.

The resistance shunt is the least expensive, but provides a small voltage and is not electrically isolated. For example, a standard metering shunt develops 50 mV at rated current. The use of a higher resistance shunt results in increased power dissipation and drift of the shunt resistance with temperature. For example, a one-volt signal at 200 A develops 200 W to be dissipated. A motor interpole winding can serve as the shunt.

A dc transductor is a form of saturable reactor operating in the high control-circuit impedance mode. The armature circuit conductor is the dc control winding, and the signal is developed in either the excitation gate winding or an additional output winding. The ac gate current will be a square wave, which will have small ripple after it is rectified. The circuit and operation of a transductor are shown in Figure 9.4. The supply

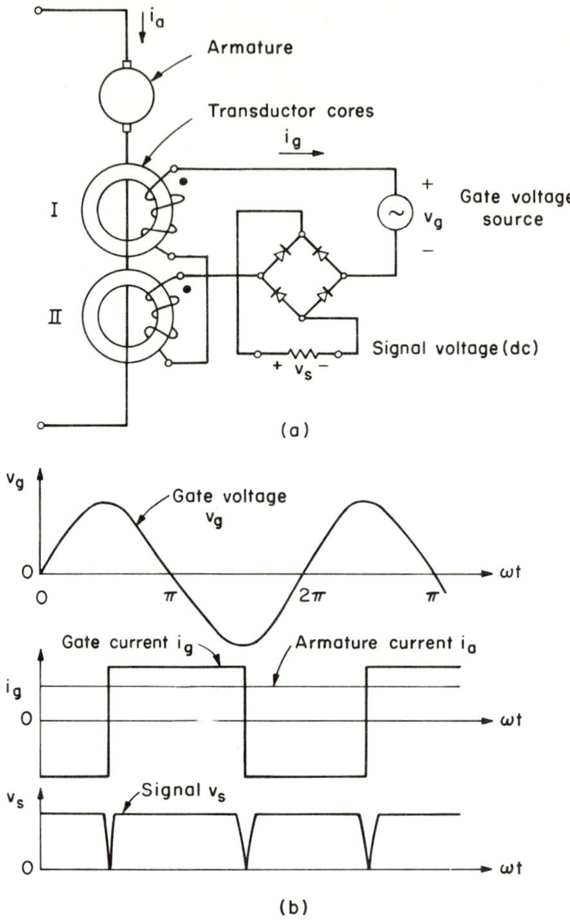

Figure 9.4   (a) Transductor current measuring circuit; (b) waveforms.

frequency can be the line frequency or higher; the accuracy is not dependent upon the frequency.

The currents in the ac lines to the thyristor circuits or bridges carry the information on the dc armature current when a freewheeling diode is not used. The rectified output of the current transformers will then yield a signal proportional to armature current. One current transformer is needed for single-phase and three for three-phase service. The circuit and waveforms for a typical three-phase installation are shown in Figure 9.5.

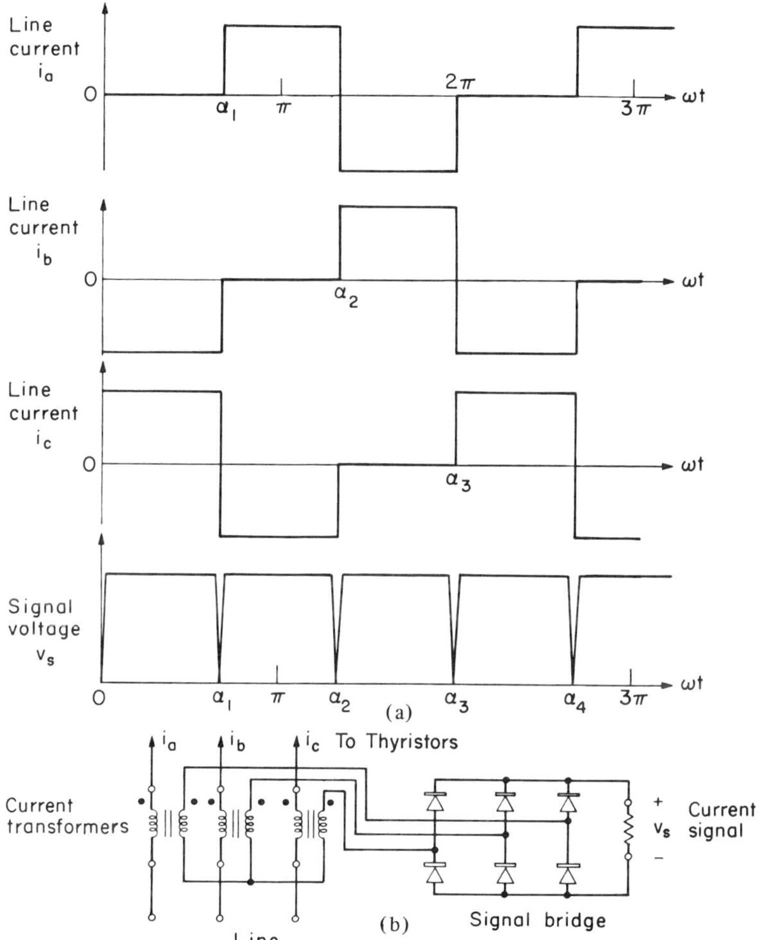

Figure 9.5   Three-phase current transformers:
(a) waveforms; (b) circuit.

9.4   *Motor Matching*[2,3,4]

Three factors govern the matching of a dc motor to a drive system. First, the rated armature voltage for base speed should match the rectified value of the line voltage so that a transformer will not be required. Second, the motor must be capable of operating with the worst-case rms armature current without overheating. Third, the motor must be able to operate without excessive sparking at the brushes.

For single-phase drives, the rectified value of 230 V is 0.9 of 230 or 208 V. However, to accommodate the low-line condition of -10 per cent and the forward drop in the rectifier elements, a motor rated 180 V is used. For 115-V full-wave service, a 90-V motor is used. For three-phase full-wave operation, the rectified value of the rms line voltage V_0 is 1.35 V_0. To accommodate the low line condition and rectifier drop a 240-V dc motor is used for a 230-V line and a 480-V motor for a 460-V line. The field windings are usually rated at half voltage. For three-phase half-wave drives, the average rectified voltage is 0.68 of the rms line voltage. For a 460-V line, a 240-V dc motor can be used, and a 120-V dc motor for a 230-V line.

The heating effect of the armature current is defined by the *form factor*, which is the ratio of the rms to average current. The form factor is a function of the firing angle of the rectifier, the armature circuit inductance and the armature generated voltage. It is also a function of whether a freewheeling diode is used. The heating is proportional to the square of the form factor. The motor must be ducted of forced cooled to compensate for the additional losses.

Curves of form factor for the single-phase full-wave rectifier are shown in Figure 9.6. The curves for Figure 9.6a for the circuit with no freewheeling diode shows that the form factor decreases as the motor is loaded and the current tends to be continuous. The form factor tends to decrease at the higher speed because the conduction period is longer for the same average armature current. Figure 9.6b shows that the free-wheeling diode has no effect on the form factor for 100 per cent speed because the instantaneous armature voltage is too high ever to go negative. However, at the lowest speed, the diode extends the conduction period of the current pulse on the trailing side and reduces the form factor. It is also obvious that the addition of a choke can more than pay for itself in reduced armature heating.

The form factor for the current for a three-phase drive is usually less than 1.11; hence, it is more convenient to measure the distortion in per cent ripple. Per cent ripple is defined as the ratio of per cent rms harmonics to average current. The form factor FF is related to per cent ripple PR by $\mathrm{FF} = [1 + (\mathrm{PR}/100)^2]^{1/2}$.

The relationship between ripple current and speed factor is shown in Figure 9.7 for a complete and incomplete three-phase bridge, and for a

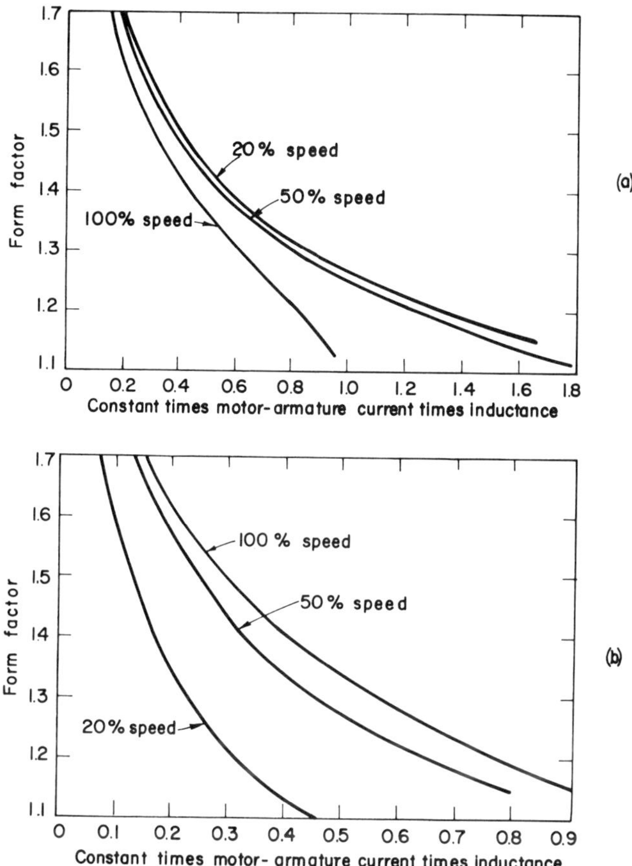

Figure 9.6   Variation of form factor as a function of motor
armature current and inductance for a single-phase
full-wave rectifier. Current flow is discontinuous. ac input
to rectifier is 230 V; dc output for 100 per cent
speed is 180 V: (a) with no freewheeling;
(b) with freewheeling rectifier.

freewheeling diode. The speed factor is defined as average voltage to peak line voltage, $E_m/\sqrt{2}\,V_0$. The two bridges have the same waveforms at full conduction, where $E_m/\sqrt{2}\,V_0 = 0.955$. As the firing angle is retarded, the per cent ripple increases, reaching a peak at about 0.5 speed factor. The incomplete bridge produces about three times the per cent ripple of the complete bridge.

Figure 9.7 shows that the per cent ripple continues to rise as the motor

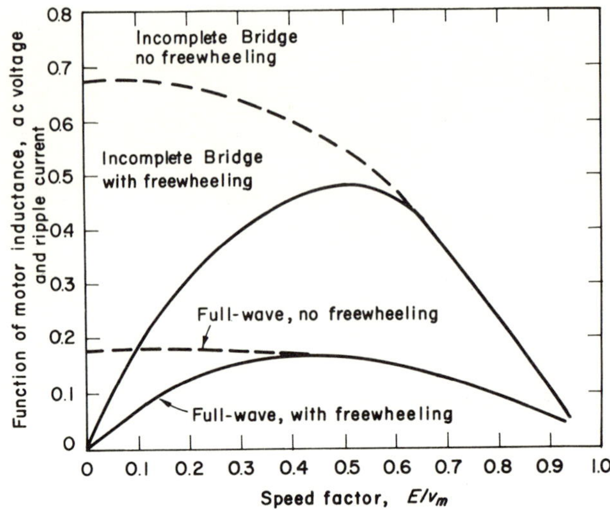

Figure 9.7   Curves showing effects of variables on ripple.

speed is reduced for no freewheeling diode. When a diode is used, the per cent ripple declines to zero at zero voltage. The half-wave bridge has the same waveform as the incomplete bridge for the lower voltages. Hence, the per cent ripple is representative of that for half-wave reversible circuits where no freewheeling diode is used. The high per cent ripple at low motor speed is troublesome because the motor cannot self-ventilate the additional losses and must be derated.

The pulsating current produced by the thyristors has a deleterious effect on commutation, which may require that a choke be added or the motor derated. A motor operating on single-phase power up to 5 hp can tolerate a form factor up to 1.35 with no commutation problems. The effect of per cent ripple on the commutation of motors operated from three-phase thyristor circuits is shown in Figure 9.8. The commutation is seen to become worse as the ripple increases and the base speed increases. The worst commutation can occur at part speed. For motors up to 150 hp, 1750 rpm, a per cent ripple up to 25 per cent can be tolerated.

9.5 Construction

The construction of solid-state drive equipment is prompted by the requirement for rapid service, the use of the same subassemblies for a variety of applications, and the demand for reliability. The designers of

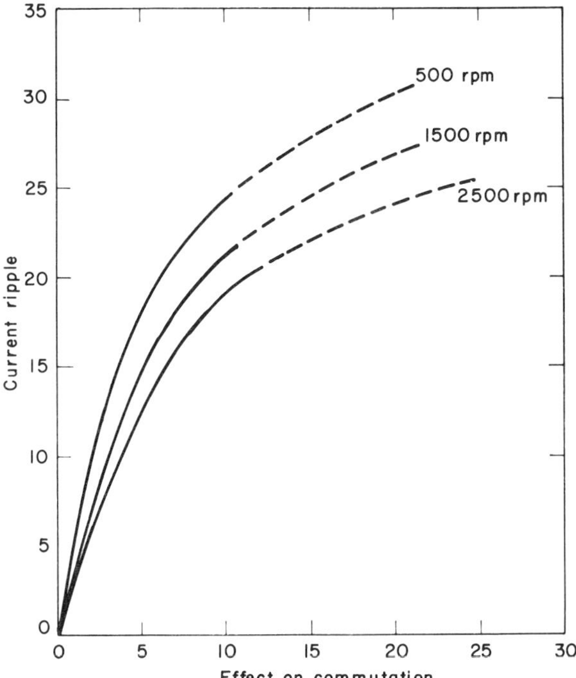

Figure 9.8 Curves showing effects of ripple on
commutation for various speeds.

early drives were concerned with electrical circuits, thyristor cooling, and other problems of a new technology. The field has now matured and designers are improving the engineering of drives.

Small single-phase drives, up to 5 hp, are built into cabinets as shown in Figure 9.9. Some of the subassemblies, such as control and firing circuits, are made plug-in to facilitate manufacture and service. The cabinets are self-cooled.

Three-phase drives are built with a high degree of modularity. The thyristors and diodes are usually assembled on heat sink assemblies as power modules. The modules require only the connection of the ac and dc power circuits and the trigger circuits. Modules are used in multiple for reversing drives and for increased horsepower ratings. Construction of a three-phase drive is shown in Figure 9.10; its diode and thyristor modules are shown in Figures 9.11, 9.12, and 9.13.

Figure 9.9 Interior of single-phase drive cabinet. (Courtesy of Electric Regulator Corp., Norwalk, Conn.)

Figure 9.10 The top row of modules contains four power diodes. The dc isolation switch is in the left-hand module of the second row, and the other three second-row modules contain power SCRs. (Courtesy of Electric Regulator Corp., Norwalk, Conn.)

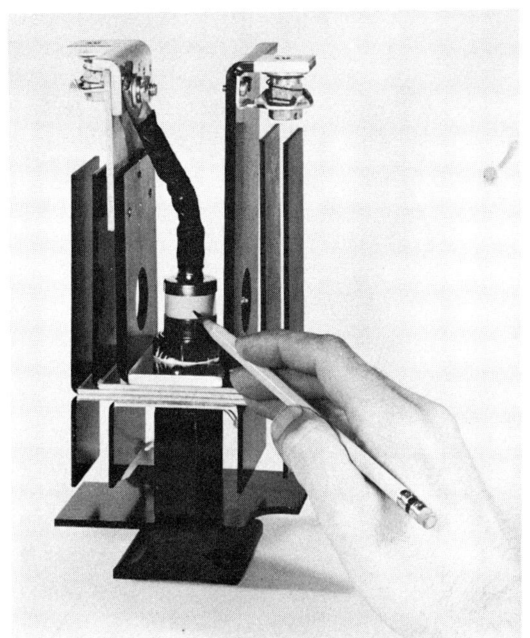

Figure 9.11 Solid-state diode module for a three-phase drive. (Courtesy of Electric Regulator Corp., Norwalk, Conn.)

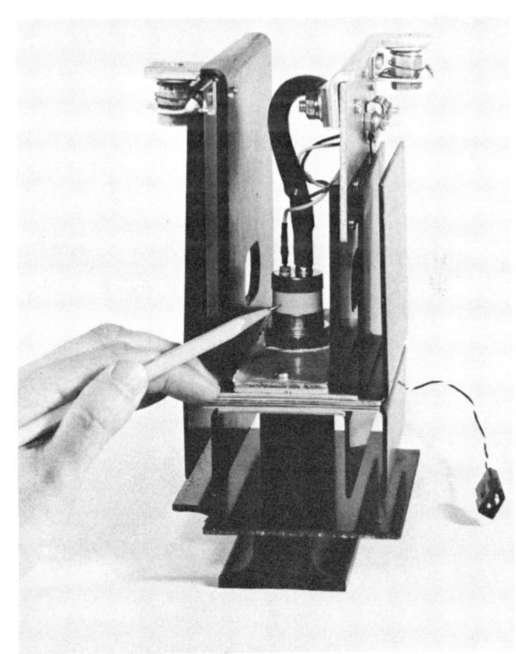

Figure 9.12 SCR power module for a three-phase drive. Fasteners, top, that mount the modular heat sink also make bus-bar power connections. Small twisted pair is the SCR gate connection, filtered by a capacitor, top right. (Courtesy of Electric Regulator Corp., Norwalk, Conn.)

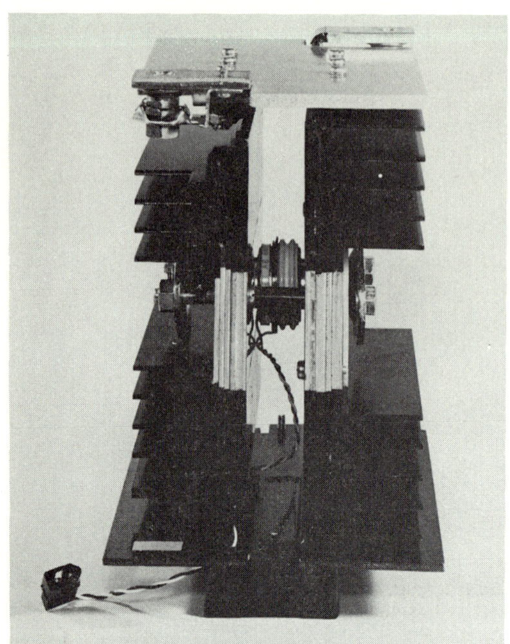

Figure 9.13   SCR power module for a large three-phase drive with a press-pack "hockey puck" SCR. (Courtesy of Electric Regulator Corp., Norwalk, Conn.)

Figure 9.14   This plug-in circuit card for a three-phase control carries magnetic amplifier trigger circuitry that generates SCR gating signals. (Courtesy of Electric Regulator Corp., Norwalk, Conn.)

Figure 9.15   Sensing and reference plug-in circuit card for a three-phase drive regulator. (Courtesy of Electric Regulator Corp., Norwalk, Conn.)

Figure 9.16   Instantaneous overload trip (IOT) relay circuitry is assembled on this plug-in regulator card. (Courtesy of Electric Regulator Corp., Norwalk, Conn.)

Control and firing circuits are prepared on separate cards by function and plugged into a card frame. Cards are selected from stock to fill the particular functions for each drive. The cards can be easily pulled out and replaced. The cards carry factory or field set adjustable potentiometers for setting up the drive system. A typical assembly of cards is shown in Figures 9.14, 9.15, and 9.16.

Most drives are designed to operate without a power transformer by matching the motor rating to the line voltage. The half-wave drives require a zig-zag grounding transformer to establish a neutral when one is not available. The transformers can be mounted within the cabinet for small drives and externally for large drives.

Ventilation is most important in solid-state drives to maintain the temperatures of control and power semiconductor equipment. Although fans will reduce the dissipating area of the heat sinks, the fans are frequently the least reliable components of the drives. Hence, the trend is to build drives for convection cooling by using massive heat sinks.

9.6 Field Winding Supply Methods

The field winding of the dc motor of an adjustable-speed drive requires a source of field current independent of the armature supply. The field current can be kept constant, or it can be reduced to extend the high-speed end of the range.

The field winding can be supplied by a separate rectifier, or it can be connected to the controlled thyristor armature bridge in such a way that current is supplied independently of firing angle of the thyristors. The field winding has a relatively large time constant, hence does not require a polyphase rectifier for low-ripple current.

The simplest form of supply is shown in Figure 9.17a. It consists of a half-wave diode D_1 and a freewheeling diode D_2. The waveform of current i_f is shown. The diode D_1 conducts when v_0 is positive; the diode D_2 freewheels the current when v_0 is negative. The shaded area represents the voltage across the inductance L_f and must be equalized over a cycle for the flux to return to its starting point. The average field current for rms line voltage V_0 is

$$I_f = \frac{0.45 V_0}{R_f}. \tag{9.2}$$

A full-wave field supply is shown in Figure 9.17b. The circuit does not require a freewheeling diode. The field current passes through the bridge arms D_1 on one half-cycle and D_2 on the second half-cycle. The line

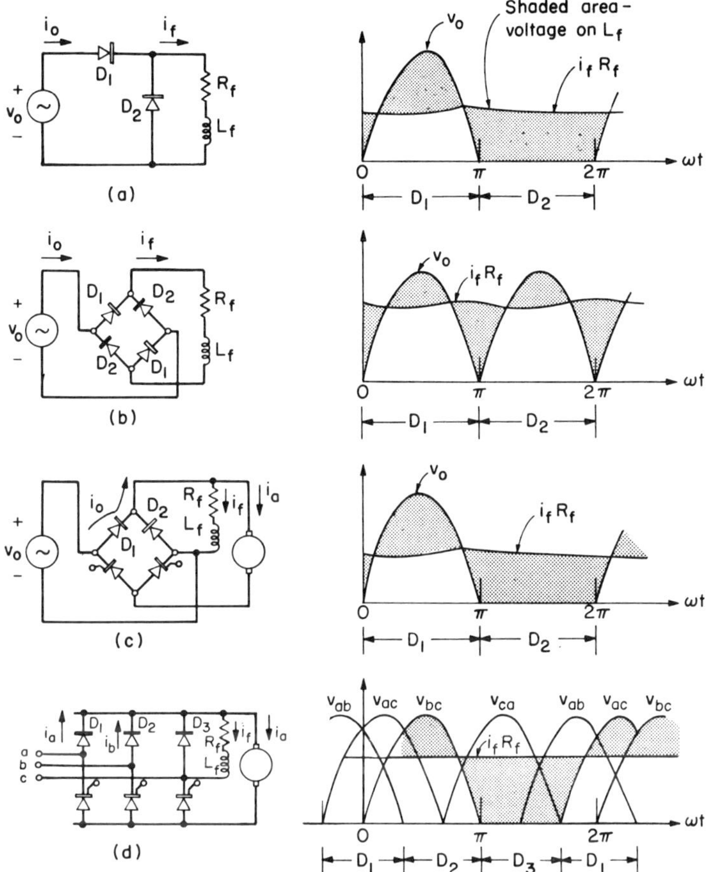

Figure 9.17   Field winding supply methods: (a) half wave;
(b) full-wave; (c) from single-phase armature bridge;
(d) from three-phase armature bridge.

current i_0 is practically a square wave. Should the ac supply be interrupted, the field current can freewheel through the bridge.

Figure 9.17c shows a single-phase incomplete bridge with two thyristors and two diodes for combined armature and field supply. The field winding is connected across the diode D_2. On the positive half-cycle of line voltage v_0, the field current flows through diode D_1, the field winding, and back to the line. On the negative half-cycle, the current freewheels through diode D_2. If the positive half-cycle thyristor is fired, then diode

D_1 carries the sum of the armature and field current. The average field current is given by Equation 9.2.

A clever method for field winding supply from a three-phase incomplete bridge is shown in Fig. 9.17d. The field winding is connected across D_3. When either voltage v_{ac} or v_{bc} is positive, current flows through the corresponding diode and through the field winding.

When neither source is available, the current freewheels through diode D_3. The average field current is given by

$$I_f = \frac{\sqrt{2}\,2V_0}{2\pi R_f} \int_0^{2\pi/3} \sin \omega t \, d(\omega t) = \frac{3V_0}{\sqrt{2}\,\pi R_f} = 0.675 \frac{V_0}{R_f}, \qquad (9.3)$$

where V_0 is the rms line-to-line voltage. The voltage $0.675\,V_0$ is one-half the dc bridge voltage at maximum conditions, so that the field winding should be rated for one-half the armature voltage for this connection.

9.7 *Speed Control By Field Weakening*

As shown in Figure 2.2, the speed of a dc motor can be raised above its base value by reducing the field magnetic flux. Such control is usually used as an adjunct to armature-voltage control of speed for specific purposes.

Control by field weakening can be carried out with just open-loop setting of the field current by an auxiliary field rheostat or a phase-controlled rectifier, with the armature at full voltage, or it can be carried out with closed-loop speed regulation. For speed regulation, a tachometer must be used because the armature voltage v_a is a measure of speed only at constant flux. A means for transfer from armature to field control must be incorporated into the speed setting dial using ganged controls and transfer relays.

The range of control by field weakening is usually limited to between 2 and 3 to 1. The maximum speed is restricted by mechanical considerations on the armature and commutator and by the instability of operation with low magnetic field flux. As the field is weakened, the armature reaction tends to distort the flux distribution causing, among other things, the motor speed to be unstable. Motors having pole-face windings can operate over wider ranges without field distortion.

The dynamic compensation of closed-loop speed control systems using the field current is more difficult than armature control. The parameters vary with flux and the armature reaction tends to produce the equivalent of positive feedback within the loop.

Speed control by field weakening is best applied where a manually controlled overspeed range is required for functions demanding less than normal torque. Typical applications are fast unwinding of reels, driving unloaded platforms to a start position, backing machine-tool bits out of the work, and so forth. Speed regulation in this range is expensive and should be used only if required.

9.8 *Protection*

Power and drive equipment utilizing magnetic and rotating machine components requires protection only against faults and overloads. However, solid-state drives combine heavy power equipment with power-handling thyristors and diodes and small signal transistors and circuit components. The protection must be extensive; it falls into two categories: protection against damage to components and protection against false operation. See Corbyn[5] for an extensive treatment of the problem. A solid-state dc drive system must be protected against the following disturbances:

1. *Line-voltage transients,* which are produced by lightning, fault-clearing and switching within the plant, and switching of transformers, particularly those supplying the drive itself. These transients can reach amplitudes of several kV on a nominal 460-V line, and produce damage by breaking down thyristors and diodes in the reverse direction, burning out small semiconductor devices, and producing false firing of thyristors. Generally, the lower the line voltage level, the less exposed is the equipment, and the lower the amplitude of the transient voltages.

2. *Overload* of the drive motor and the power semiconductors. The overload is manifested as overcurrent and can result from sustained overload, acceleration and reversing, and operation at low speed with poor ventilation. The semiconductors have very small thermal storage capability and are particularly vulnerable to overcurrent conditions that excessively raise the junction temperature.

3. *Faults* in the drive motor or the thyristor bridges. The most serious fault in the motor is a commutator flashover which places a dead short circuit across the dc bus. The most common faults of the thyristor bridges are permanent reverse breakdown of an element, or *shoot-through*, temporary forward conduction of a thyristor. These faults appear as short circuits across the ac phases and/or short circuits across the dc bus. The danger of the latter is that the motor can act as a source of fault current and feed a dc bus fault even though the ac line is interrupted. The fault currents must be interrupted before the semiconductor devices are damaged.

4. *Noise,* which is generated by the switching of currents in inductive circuits within and external to the drive system. The noise appearing on the lines to the thyristor gate terminals can cause false firing and produce faults in the thyristor bridges.

5. *High dv/dt and di/dt* on the thyristors resulting in false firing for high dv/dt, when the thyristor is supposed to be blocking, and high junction temperatures for high di/dt. The suppression of these disturbances is part of the design problem of insuring reliable usage of the drive system at its maximum ratings.

Several examples of complete systems with their protective devices are shown in earlier chapters, and in the references. The protective devices are described as follows:

1. *Thyrectors* are selenium rectifiers operated in the reverse direction to act as voltage-sensitive nonlinear resistors. Each cell has a break over voltage of about 30 V, the cells are stacked in series to obtain the desired level. The plate area per cell determines the energy that the cell can dissipate. The thyrectors can be assembled back-to-back to operate on alternating voltage. The thyrectors are suitable for clipping voltage transients that contain high energy. They are usually placed across the line terminals and across the armature terminals. They are not suitable for fast transients and allow a significant increase of voltage to carry heavy transient currents.

2. *Capacitors* are used across the secondary windings of transformers to suppress transient voltages, particularly those produced by interruption of the magnetizing current. Such capacitors are in the microfarad range.

3. *Fuses* are used to protect solid-state devices. They are usually placed in the ac lines to the thyristor bridges and selected to interrupt the current before the bridge element fails. The fuses are either fast fuses or conventional fuses. The first type can interrupt the current very rapidly so that the thyristors and diodes can be operated without large current margin. The conventional fuses are less expensive and are always available for replacement. Fuses are selected by coordinating the I^2t of the fuse with that of the thyristor.

4. *A dc circuit breaker* is placed in the armature circuit of larger horsepower drives and operated from a trip circuit. The breaker trips in case of motor fault, overload, backfeed of the motor to the bridge.

5. *RC circuits,* called snubbers, are used across the ac lines to suppress fast transients, across the thyristors and diodes to damp high voltage transient "spikes," and to limit the rate of voltage rise (dv/dt). Typical values are $R = 50\ \Omega$ and $C = 0.1\ \mu F$, with a time constant of 5 μs.

6. *Shielding and small capacitors* are used for the gate signal lines to suppress the pickup of voltage transients that could cause false triggering. The capacitors are less than $0.1\ \mu F$ and must not deteriorate the waveshape of the pulse, nor deliver too large a current into the gate terminal when the thyristor fires.

The degree of protection required rises with voltage level and current level. A fractional-horsepower drive may require only a fuse and an *RC* snubber. A 100-hp drive requires a sophisticated protection system acting on the sources of disturbances and the points of impact.

In addition to the protective devices, the semiconductors themselves are usually derated, typically by factors of two-to-three on voltage and current.

References

1. A. V. Wise and R. G. Schlieman, "A dc motor drive using the new pressure assembled thyristors," *Proceedings of the Second IGA Conference,* pp. 71–79, 1967.
2. C. J. Newell, "Matching d-c drives and motors," *Electrotechnol.,* vol. 79, No. 6, pp. 38–41, June 1967.
3. N. Kaufman, "An application guide for use of d-c motors on rectified power," *Trans. IEEE, Power App. Systems,* pp. 1006–1009, October 1964.
4. C. E. Robinson, "Redesign of dc motors for applications with thyristor power supplies," *IEEE Trans. Ind. and Gen. Appl.,* vol. IGA-4, no. 5, pp. 508–514, September/October 1968.
5. D. B. Corbyn, "Voltage surge control in thyristor equipment," *IEE Conference on Power Applications of Controllable Semiconductor Devices,* no. 17, pt. 1, pp. 89–101, November 1965.

TABLE OF SYMBOLS

B	mechanical damping
C	capacitance
G	conductance
i_a, I_a	armature current; instant, dc, average
i_f, I_f	field current; instant, dc, average
i_m	series motor current, instant
i_m, I_m	load current, instant, rms, average
i_a, i_b, i_c	three-phase line currents, instant
J	inertia
K_a	armature voltage constant
K_m	magnetic field factor
K_t	armature torque constant
k_1	analog scale factor
k_r	pulse ratio
L_a	armature inductance
L_f	field inductance
N	speed
R_a	armature circuit resistance
R_f	field circuit resistance
R_c	control resistance
R_n	load resistance
t	time
T	torque
T_a	armature time constant
T_m	mechanical time constant
v_a, V_a	armature voltage; instant, dc, average
v_f	field voltage, instant
v_m, V_m	armature circuit voltage, instant, dc
v_n, V_n	load voltage; instant, rms, average
v_0, V_0	line voltage; instant, rms
v_{an}, v_{bm}, v_{cn}	line-to-neutral voltages, instant
v_{ab}, v_{bc}, v_{ca}	line-to-line voltages, instant
X	reactance
α	firing angle
β	extinction angle
γ	conduction angle
Φ_f	field magnetic flux
ϕ	impedance angle
θ	shaft angle
τ	time constant
ω	angular velocity

INDEX

Base speed, definition, 95

Chopper circuits, 84, 86
  regeneration, 89
Circuit diagrams, drives
  commercial single-phase, 56
  commercial three-phase, 67
Commutating thyristor, 87
Commutation, dc motors, 24
  effect of ripple, 24
Construction, 110
Current limit, definition, 97
Current sensing
  current transformers, 105
  shunt, 105
  transductor, 105

Drives, single-phase
  full-wave, 48
  Gaudet, 51
  half-wave, 44
Drives, three-phase
  complete bridge, 61
  dynamics, 64
  half-wave, 64
  incomplete bridge, 58
  reversing, 64
Drives, types
  adjustable speed, 1
  constant horsepower, 2
  series universal motor, 4
Dynamic braking, definition, 96

Field winding supply, 116
Firing angle, voltage, curves
  single-phase, half-wave, 30
  three-phase, bridge, 43
  three-phase, bridge, inductive
    load, 63
  three-phase, half-wave, 40
Firing circuits
  magnetic, 99
  solid-state, 101
  three-phase, 103
Form factor, 108
Freewheeling, 49

IR-drop compensation, definition, 95

Jones chopper, 86

Modularity, construction, 111

Phase control rectification,
    definition, 98
Protection, 119

Rectifiers
  conduction angle, 45
  discontinuous conduction, 39, 50, 53
  extinction angle, 32
  firing angle, 28
  single-phase, reactive load, 31
  single-phase, resistance load, 28
  three-phase, resistance load, 37
  volt-time area, inductance, 32
Regenerative braking, definition, 96

Series motor, 16
  resistance control, 18
  saturation, 19
  voltage control, 17
Series universal motor drives
  full-wave, 70, 71
  Gutzwiller, 70, 72
  half-wave, 70, 71
  Momberg, 74
  phase control, saturated, 79
  phase control, unsaturated, 76
Shunt motor, 10
  analog, 15
  block diagram, 14
  commutation, 24
  dynamics, 12
  parameters, 20
  speed control, 11
Speed control by field weakening, 118
Speed range, definition, 94
Speed regulation, definition, 93
Speed sensing
  armature voltage, 104
  tachometer, 105

Torque limit, definition, 96

Torque-speed diagrams, 6
  adjustable speed drive, 2
  constant horsepower, 3
  Momberg, series motor, 76
  saturated series motor, 19
  series motor, half-way thyristor, 81, 82
  series motor, resistance control, 18
  series motor, voltage control, 17
  single-phase, full-wave, 54